Global Institute for Sustainable Prosperity

Nikolaos Karagiannis · John E. King
Editors

Visions and Strategies for a Sustainable Economy

Theoretical and Policy Alternatives

palgrave
macmillan

Editors
Nikolaos Karagiannis
Department of Accounting,
Economics and Finance
Winston-Salem State University
Winston-Salem, NC, USA

John E. King
Greensborough, VIC, Australia

ISSN 2730-5511 ISSN 2730-552X (electronic)
Global Institute for Sustainable Prosperity
ISBN 978-3-031-06492-0 ISBN 978-3-031-06493-7 (eBook)
https://doi.org/10.1007/978-3-031-06493-7

This Palgrave Macmillan imprint is published by the registered company Springer Nature Switzerland AG
The registered company address is: Gewerbestrasse 11, 6330 Cham, Switzerland

In memory of my wife Mary Nicholls
—John Edward King

In memory of my late father Vasileios Karagiannis
—Nikolaos Karagiannis

Visions and Strategies for a Sustainable Economy: Theoretical and Policy Alternatives

Setting the Context

In the current age, the issue of environmental sustainability represents one of the most critical human concerns that has drawn global attention, generated worldwide discourse, and engaged a large number of international forums and conventions, national institutions, and non-governmental organizations (NGOs). An environmentally sustainable mode of economic development has gained increasing prominence due to disastrous events and issues such as nuclear accidents, oil spills, water pollution, deforestation, the greenhouse effect, rise in sea levels, ozone depletion, land degradation, and a loss of biodiversity. Seeking to link and prioritize among aspirations pertaining to human welfare, the "sustainable development" notion stresses the long-term compatibility of the economic, environmental, and social dimensions of development, while acknowledging possible competition across these areas in the shorter term.

Despite an extensive amount of literature on "sustainable development", it still has not been possible to come up with a consensus definition of such a complicated concept. Instead, there are two distinctive perspectives on sustainable development: one supporting the continuation of economic growth through environment-friendly activities, and another stressing the reduction of economic growth and consumption for greater sustainability. In examining major factors causing or worsening the various forms of environmental unsustainability, few are briefly spelled out: a drastic increase in global population; the danger presented by poverty and inequality to the environment; excessive urbanization and industrial expansion; the rapid demise of social and welfare programmes and the rise of market-led policies; and the global expansion of unrestrained market forces.

However, orthodox analyses and proposals tend to be too economistic in outlook in terms of their tendency to convert all issues related to the environment and sustainability, including the moral questions, into economic parameters. Consequently, they fail to take into account some of the most basic characteristics of how *human* societies and economics function and develop. For instance, they largely ignore the specific dynamics of pioneer development, power imbalance in and between human societies, the role of conflict, and the fundamental diversity of interests and lifestyles. Most importantly, such orthodox studies define life-support systems almost exclusively in bio-geophysical terms, thus ignoring the fact that human progress primarily depends on the accumulated scientific and technological knowledge and on the cultural heritage of institutions and arrangements, which represent successful solutions of social, economic, and political problems. In short, orthodox contributions have failed to capture the array of real issues and policy dilemmas.

Given these limitations of mainstream approaches, it is essential that alternative perspectives on sustainability gain momentum towards providing a more comprehensive understanding of the environmental, economic, financial, educational, and civic traits of individuals, communities, and regions around the world. In this regard, there is also the transdisciplinary effort of "ecological economics", which seeks to link ecology, from the natural sciences, and economics, from the social sciences. Its goal is to develop a deeper understanding of the complex relationships between humans and nature, and to use that understanding to develop policies that will lead to a world which is ecologically sustainable, has a fair distribution of existing resources, and efficiently allocates resources, including "natural" and "social" capital.

Such an ambition clearly requires novel and fresh approaches that are comprehensive, adaptive, integrative, multiscale, pluralistic, and evolutionary, and which acknowledge the enormous uncertainties involved. Ultimately, the challenge to achieving sustainability is what constitutes the "good life" and a "better society" in the twenty-first century, and why finding ways other than simply reducing material impacts on the natural environment is part of achieving this goal. That, in turn, requires much more attention to policy interventions and less to markets. As the pace and scale of local sustainability efforts grow—and there is considerable evidence to show that they are growing—social subjectivities and collective consciousness that must go hand-in-hand with technological innovation and material change may spread. Whether this will, eventually, bring us to sustainability is something we can only wait to see.

The Structure of the Book

The multidisciplinary edited book *Visions and Strategies for a Sustainable Economy: Theoretical and Policy Alternatives* provides a thorough examination at the theoretical and, especially, policy levels of a number of key topics related to a sustainable economy and a better society. With important contributions by distinguished academics, the book presents alternative views, provides an assessment of contemporary realities in an era of ecological emergency, and offers visions, strategies, and realistic policies towards a better economy and society while paying special attention to a 'green new deal' for different areas.

In Chapter 1, Dale provides an analysis of sustainability, self-sufficiency, and sufficiency in the ancient world. According to the author, the sustainability discourse is thoroughly modern, and so too are strategies to achieve it via "sustainable consumption" and "sufficiency". Considering that consumerism should be understood as an ideology organic to capitalism, Dale proposes that sufficiency needs to appear in a radical-democratic light as a demand for an "alternative hedonism" or, broadly conceived, as class struggle.

In Chapter 2, Arestis, Karagiannis and Madjd-Sadjadi link the developmental state view to green economic growth. The authors briefly review the developmental state literature and point out the absence of "green considerations". To address this missing link, Arestis, Karagiannis and Madjd-Sadjadi offer a green developmental state framework by distinguishing two main sets of factors: sociocultural and politico-institutional elements, and economic development and industrial targeting aspects.

In Chapter 3, Sawyer develops the case for an environmental industrial strategy, which requires a comprehensive approach by government to industry. The author then examines the ways in which macroeconomic policies can be used to support the environmental industrial strategy with regard to fiscal policies and to monetary policies, as well as the ways in which the investment required under such a strategy could be secured to ensure that there is a successful transition to a sustainable economy.

In Chapter 4, Arestis and Karagiannis focus on macroeconomic policies employed during the Global Financial Crisis and Covid-19 and, particularly, on future economic policies which are deemed necessary to achieve high levels of economic activity and employment along with equal distribution of income and wealth. According to the authors, these future economic policies need to be properly coordinated if successful economic activity is to be achieved. However, as Arestis and Karagiannis contend, fiscal and monetary authorities should do their best to keep such coordination safe, resilient and transparent.

In Chapter 5, Rossi takes the flaws of neoliberalism into account to contend that central banks should not limit their policy focus to price stability but must, first and foremost, act to guarantee financial stability –which has become a major issue after the outbreak of the global financial crisis in 2008 and the Covid-19 pandemic. The author further argues that monetary policy should contribute to a collective effort in supporting and carrying out an ecological transition that favours economic development for the common good. To achieve this goal, governments can issue "green bonds" that central banks should purchase to finance the ecological transition, to be achieved also through an appropriate interest rate policy.

In Chapter 6, Correa updates the ruminations of Keynes by developing a model in which government investment alone and people without work are connected by a long-lived government instrument that is held by people and provides them with returns. For this connection to be successfully accomplished, governments will have to ensure domains where their interests are aligned with the people; but pursuing self-interest is best served by policy coordination.

In Chapter 7, Nersisyan and Wray set out a framework for analyzing the true cost of the Green New Deal (GND) in the United States. The authors change the debate from estimating financial costs to a careful assessment of resource needs and availability. Such an approach, informed by Keynes and Modern Money Theory (MMT), provides more useful guidance about the question of the GND's "affordability" than the conventional method of merely adding up the "dollar costs" of GND projects.

In Chapter 8, Messkoub proposes a growth or degrowth strategy that can manage/negotiate the composition of output whilst investing in resources to reduce depletion of natural assets and greenhouse emission. Such a strategy is based on the "human theory of needs" that not only meets the needs of the current generation but also provides some measure of inter-generational justice. The welfare and social policy counterpart of this approach must involve public and collective provisioning of socially necessary services of health, education, and a range of other care services that can reduce per capita cost through economies of scale and scope whilst providing an equitable access to these services.

In Chapter 9, Bortis utilizes the term "environmental alienation" to sum up the overall deviation from an ideal sustainable economy, which is one that implies harmony between society and nature. The author's argument focuses on an additional impetus to be given at moving towards a sustainable economy through implementing, step by step, social liberalism as a new paradigm, which is desperately needed to permanently bring about a sustainable economy.

In Chapter 10, Lewney explores pathways by which the increase in global temperature relative to pre-industrial levels might be kept 'below 2° C'. The author discusses the role that Green New Deal policies can play in stimulating the transition, which may also promote a Green Recovery from the impact of the Covid-19 pandemic. The discussion highlights the importance of a multi-disciplinary approach to modelling as well as the implementation of key insights of Schumpeterian and post-Keynesian economics in a detailed empirical framework.

In Chapter 11, Barkin proposes a different epistemology and methodology reflecting the direct participation of a wide diversity of communities around the world, particularly in the Global South, as well as their possibilities for implementing different approaches in order to improve their well-being. The author focuses on the emergence of alternative social and productive structures that respond to the demands for local control of the governance process in order to assure local well-being, responsible environmental management, and a balanced ecosystem as intimately related themes of social justice.

In Chapter 12, Karagiannis and Mohammed extract relevant aspects of the East Asian successes and embed them in an appropriate framework that emphasizes "green" and "blue" growth strategies for Caribbean economies. Taking current multifaceted impediments into consideration, the authors propose a developmental state argument as a pragmatic alternative towards addressing self-sustained growth challenges of Caribbean nations, and offer concrete policy considerations which are deemed necessary to boosting Caribbean industries' "green" and "blue" growth potential.

In Chapter 13, Huang examines China's environmental reality and the evolution of its environmental policy since 1978, which has evolved from contradiction to synthesis. The author argues that the Chinese state has been, and will continue shaping the country's environmental landscape more responsively and effectively now and into the future. In Huang's analysis, China's environmental policy today is in a historically stronger political and institutional position to take advantage of creative public policies to engineer environmental sustainability while, at the same time, achieving economic growth and full employment.

Our task would have been incomplete if we had not acknowledged those who kindly aided us in completing this book. We wish to thank the advisors and staff of Palgrave Macmillan who have provided excellent support throughout the preparation of this manuscript, and the distinguished contributors for their willingness to participate and respond to our comments and suggestions. We would also like to thank the Levy Economics Institute of Bard College and the Global Institute for Sustainable Prosperity for kindly permitting shorter versions of the Working Papers No. 931 (May 2019) and No. 112 (December 2015) to appear as Chapters 7 and 11, respectively, in this edited volume. Last, but not least, we thank our families for their continuous support and encouragement. We owe them more than we can recount.

December 2021

Nikolaos Karagiannis
John E. King

Contents

Editors and Contributors

About the Editors

Nikolaos Karagiannis is a Professor of Economics at Winston-Salem State University, North Carolina; a Research Scholar at the Global Institute for Sustainable Prosperity; an invited visiting scholar at the University of Cambridge, England; and the co-editor of *American Review of Political Economy*. He has published extensively in the areas of economic development, public sector economics, and macroeconomic policy analysis. He is the author, co-author, and co-editor of twenty-three books and has published over 160 papers as refereed journal articles, book chapters, and op-eds.

John E. King retired from teaching at La Trobe University, Melbourne, where he is now Professor Emeritus. His continuing research interests are in the history of heterodox economic thought, in particular Marxian political economy and Post-Keynesian economics. King has published over 110 papers in refereed journals and over 85 book chapters, and is the author, co-author, or editor of over 30 books. He is, and has been, a member of editorial boards and a referee for a number of economics journals.

Contributors

Arestis Philip is a Professor and Director of Research at the Cambridge Centre for Economics and Public Policy, Department of Land Economy, University of Cambridge, UK, and Professor of Economics at the Department of Applied Economics V, University of the Basque Country, Bilbao, Spain. He served as a Chief Academic Adviser to the UK Government Economic Service and as a consultant on the Central Asia Regional Economic Cooperation programme. Arestis is the holder of numerous recognitions, honours, and awards for his dedication and lifetime contributions to Economics. He has published as sole author or editor, as well as co-author and co-editor, a number of books, contributed chapters to numerous books, produced research reports for research institutes, and has published widely in academic journals. He is, and has been, on the editorial board of a number of economics journals.

Barkin David holds a Ph.D. in Economics from Yale University and is a Distinguished Professor at the Universidad Autónoma Metropolitana, Xochimilco Campus in Mexico City. He collaborated in the founding of the Ecodevelopment Centre in 1974, and received the National Prize for Political Economy in 1979. He is a member of the Mexican Academy of Sciences and an Emeritus member of the National Research Council. He collaborates with indigenous and peasant communities to promote the sustainable management of regional resources and alternatives to development towards a world of "good living". He is recognized for his theory of Radical Ecological Economics, developed during the past 15 years. In 2016, he received an award from the Alexander von Humboldt Foundation (Germany). His latest books are: *From Protest to Proposal: 50 years imagining and building the future* (Siglo XXI, 2018) and *The Environmental Tragedy in Latin America and the Caribbean* (ECLAC, 2020, with 20 Latin American colleagues).

Bortis Heinrich is a Professor of Emeritus at the University of Fribourg, Switzerland, since 2015, where he was a Professor of Political Economy from 1980. He has a Ph.D. in Economics from Cambridge (Churchill College, 1977). His supervisors were Phyllis Deane and Nicholas Kaldor (thesis supervisor). Since 1998 he has been a Life Member of Clare Hall, Cambridge University and, since 2013, a member of the German Keynes Society. His work is on fundamental issues in economic theory, specifically on the synthesis of Keynesian and Classical Political Economy. Publications include: *Institutions, Behaviour, and Economic Theory—A Contribution to Classical-Keynesian Political Economy*, Cambridge University Press (1997), *Keynes and the Classics—Notes on the Monetary Theory of Production, in Modern Theories of Money*, Edward Elgar (2003), and Post-Keynesian Principles and Policies, in *Oxford Handbook of Post-Keynesian Economics* (2013).

Correa Romar was the Reserve Bank of India Professor of Monetary Economics, University of Mumbai, till November 2021. In 2004, he was a Visitor, Maison Des Sciences de L'Homme, Paris; French Government Post-Doctoral Scholar, 1996–1997, Groupe de Réchèrche sur la Regulation de L'Economie Capitaliste, Université Pierre Mendes, France; Visiting Senior Research Fellow, Institute of South Asian Studies, National University of Singapore, 2008. He was a Member, Working Group on *Money Supply: Analytics and Methodology of Compilation*, Reserve Bank of India, 1998. He has published in *Keio Economic Papers*, *Journal of Economic Integration*, *History of Economic Ideas*, *American Review of Political Economy*, *International Review of Applied Economics*, *International Game Theory Review*, *Evolutionary and Institutional Economics Review*, *International Journal of Political Economy*, *Applied Economics Letters*, *Economics e-journal*. He has co-authored a book with Amelia Correa, *Stock-Flow-Consistent Models and Institutional Variety*, 2017.

Dale Gareth teaches politics at Brunel University. He has written several books on Karl Polanyi and several on East Germany, and has edited collections on green growth, revolutionary uprisings, Eastern Europe, and international migration. His essays have appeared in *The Ecologist, The Conversation, Truthout, Open Democracy, Viewpoint, Jacobin, and Spectre.*

Huang Vincent is a Heterodox Economist with the primary fields of political economy of China and East Asia, money and banking, and ecological economics. He is a Teaching Assistant Professor of Economics at the University of Denver and a Research Scholar at the Global Institute for Sustainable Prosperity. Vincent received his Interdisciplinary Ph.D. degree in Economics and Social Science Consortium at the University of Missouri–Kansas City in 2020.

Lewney Richard is a Chair of Cambridge Econometrics (CE), a consultancy spin-off from the University of Cambridge, and a Trustee of CE's controlling shareholder—the Cambridge Trust for New Thinking in Economics. CE's multiregional, macrosectoral model, E3ME, is one of the world's leading applied energy–economy–environment models. He has directed major research projects to improve the methods of modelling the macroeconomic impacts of low-carbon policies, and a range of projects in which that modelling is applied to assess particular economic and technology policies.

Madjd-Sadjadi Zagros is a Professor of Economics and the Chair of the Department of Accounting, Economics and Finance at Winston-Salem State University, NC, USA; the former Chief Economist of the City and County of San Francisco; and the founding editor of the *American Review of Political Economy*. He is a former Fulbright Scholar to Queen's University, Canada and has written dozens of academic articles and book chapters, authored two textbooks in the field of law and economics, and was the co-author with Nikolaos Karagiannis of *Modern State Intervention in the Era of Globalization* (2007).

Messkoub Mahmood is currently Senior Research Fellow at the International Institute of Social Studies (ISS, Erasmus University of Rotterdam, NL) working with a consortium of EU and African universities and institutions on an EU (Horizon 2020) funded research project on the narratives of migration. His other research interests are in the areas of social policy, economics of population ageing and universal approach to social provisioning. His recent publications are related to social policy and population ageing in MENA, poverty and employment policies in MENA, unpaid household work and cash transfers in Iran and EU Social Investment. He has acted as a consultant to ESCWA, ILO and the UN (DESA, UNFPA). He researched and taught economics, development studies, social policy and population studies (mobility/migration, age structure and ageing) at universities of London (Queen Mary), Leeds, and Erasmus (ISS).

Mohammed Debbie A. obtained her Ph.D. from the University of the West Indies in 2005. She is currently Professor of International and Global Business at Centennial College in Toronto, Canada. She was a Senior Lecturer in International Trade at the Institute of International Relations and the Arthur Lok Jack Graduate School of Business, the University of the West Indies, Trinidad and Tobago, and one of 5 members of the EU-LAC Academic Council. She is the author of two books and has published extensively in academic journals and contributed book chapters to several important scholarly works in the areas of Caribbean economic development, competitiveness, regional economic integration, and trade facilitation. Her current research focuses on culture and governance as transformative factors for socioeconomic development.

Nersisyan Yeva is an Associate Professor of Economics at Franklin and Marshall College in Lancaster, PA. She received her B.A. in Economics from Yerevan State University in Armenia, and her M.A. and Ph.D. (2013) in Economics and Mathematics from the University of Missouri–Kansas City. She is a macroeconomist working in the Modern Money Theory, Post-Keynesian, and Institutionalist traditions. Her research interests include banking and financial instability, fiscal and monetary theory, and policy. She has published a number of papers on the topics of shadow banking, liquidity creation, the Glass-Steagall Act, the Green New Deal, and Modern Money Theory. Currently, she is working on editing the Elgar Companion to *Modern Money Theory* with L. Randall Wray.

Rossi Sergio is a Full Professor of Economics at the University of Fribourg, Switzerland, where he has held the Chair of Macroeconomics and Monetary Economics since 2005. His research interests are in macroeconomic analysis, particularly as regards national as well as international monetary and financial issues. He has authored or edited about 25 books including an encyclopaedia of central banking and an encyclopaedia of post-Keynesian economics, has widely published in academic journals, and is frequently invited to TV talk shows discussing contemporary macroeconomic issues both at national and international levels. He is a member of the editorial boards of *Cogent Economics and Finance*, *International Journal of Monetary Economics and Finance*, and *Review of Political Economy*. Since 2015, he has been featuring in the list of the most prominent economists in Switzerland published by the *Neue Zürcher Zeitung* yearly.

Sawyer Malcolm is an Emeritus Professor of Economics at the University of Leeds, UK. He was the lead coordinator for the EU-funded €8 million five-year project on "Financialisation, Economy, Society and Sustainable Development" (www.fessud.eu). He established and was managing editor of the *International Review of Applied Economics* for three decades. He is the editor of the book series *New Directions in Modern Economics* (Edward Elgar) and coedits (with Philip Arestis) the annual series *International Papers in Political Economy* (Palgrave). He is the author of 15 books and editor of over 30 books. He has also published over 100 papers in refereed journals and over 100 book chapters on a wide range of topics, recently including writings on financialization, fiscal policies, ecological macroeconomics, political economy of the euro, analysis of money, and the nature of macroeconomics.

Wray L. Randall is a Senior Scholar at the Levy Economics Institute of Bard College, Annandale-on-Hudson, NY and Professor of Economics at the University of Missouri–Kansas City, MO, USA. He is currently working in the areas of monetary policy, employment, and social security. Wray has published widely in academic journals and is the author or editor of several books on endogenour money, modern money theory, J.M. Keynes, Minsky, and A.M. Innes. Wray holds a BA from the University of the Pacific and an MA and a Ph.D. from Washington University in St. Louis, USA.

List of Figures

Green New Deal Policies and the Decarbonisation Challenge

China's Ecological Civilization: From Contradiction to Synthesis

List of Tables

Sustainability and Development Policies

Sustainability in the Ancient World: Sufficiency as a Strategy of Aristocratic Hegemony

Gareth Dale

1 Introduction

"Sustainability," since the 1990s a touchstone of liberal politics, has lost meaning in proportion as it has gained currency. It has become semantic flotsam, unmoored from any settled sense. In this paper I shall constrain it to one of its meanings: the recognition that human flourishing, locally or globally, requires living within ecological constraints, with resource and energy use and pollution kept below levels that cause serious environmental harm. This goal is normally discussed in supply-side terms, centred on the redesign of production and distribution systems to reduce material and energy inputs ("efficiency") or through demand reduction ("sufficiency"). My own view is that both directions are needed, and more. Demand reduction, naturally, should centre on the rich world and especially among the richest 1%. They, according to the UN Environment Programme (UNEP 2020), need to shrink their carbon emissions "by at

G. Dale (✉)
Brunel University, London, UK
e-mail: Gareth.Dale@brunel.ac.uk

N. Karagiannis and J. E. King (eds.), *Visions and Strategies for a Sustainable Economy*, Global Institute for Sustainable Prosperity,
https://doi.org/10.1007/978-3-031-06493-7_1

least a factor of 30" if the 1.5 °C goal of the Paris Agreement is to be achieved. But how should the case for sufficiency be framed?

In this chapter I approach the question through a study of sustainability, self-sufficiency and sufficiency in the ancient world. Ancient civilisations have long inspired anti-consumerist and anti-capitalist critiques. Some look to Daoism, with its creed of frugality, localism and communion with nature, and its denigration of greed and infinite desire, or to Jainism, with its compassion for all animals and plants. Some look to Christian monasticism and the early church fathers, to their rejection of excessive wealth accumulation on the grounds that it undermines psychological balance and distracts from devotion to God (Ambrose 1961, p. 267). Still others look to Ancient Greece and Rome. Here the guide is Rousseau, who contrasted the peoples of Antiquity, following simple, virtuous lives close to nature, with modernity, its commercial economy promoting inequalities, corruption and greed. The Rousseauian lineage persists today, in conservative and radical versions. Representative of the former is the political theorist William Ophuls. Sustainability, he argues in *Plato's Revenge* (2011, p. xi), is an "oxymoron" in our industrial society, which resembles the Titanic and cannot be "greened." We should instead trade industrial society for an ascetic, militarised, simpler and "humbler vessel"—and one whose captain won't bow to the desires of passengers or crew. Drawing on Plato and Le Bon, alongside Rousseau, he makes the case that achieving simplicity and frugality will require a restraining of the masses' passions by a wise elite—a "natural aristocracy" of scientist-technocrats who "can supply the discipline, prudence, and forethought necessary to check the destructive tendencies of crowds" Ophuls (1992; 2011, p. 150, and in Quilley 2013, 2017). Conversely, the elite's power must itself be checked, ideally on lines advanced by thinkers of the Stoic tradition, the Hellenistic school of philosophy that was enjoying its second wind in the Roman Empire at the time of Seneca (c. 4 BCE–CE 65). The political philosophy of Seneca and Cicero is suited to the ecological ethos we moderns need, Ophuls maintains (2011, p. 184). In finding his muse among the Stoics, Ophuls is far from alone. The Stoics' ethos of prudence and self-control is "intrinsically linked to sustainable well-being," Kai Whiting and his colleagues (2018) suggest. It should assist us today in designing a new social order, capable of flourishing without "limitless growth."

Among radicals, too, the Rousseauian lineage is alive. As an illustration, consider the argument of eco-feminist Vandana Shiva (2013).

"Economics" and "ecology," she reminds us, are both derived from *oikos*, the Greek for household—a term that encompasses a plot of land, its buildings, its animals and people. From this she infers that in the age when "economics was focused on the household, it recognized and respected its basis in natural resources and the limits of ecological renewal," organising the provision of "basic human needs within these limits." *Oikos* economics, she adds, was "women-centred." On similar lines, the Green Party MEP Molly Scott Cato (2009, p. 18) commends Aristotle for the distinction he draws between the authentic economy, *oikonimia*, which deals with "social and natural resources" and its offshoot, *chrematistics*, the art of acquiring goods, which promotes greed and obsessions with "property, wealth and currency," and, relatedly, for his acknowledgement of the "key green principle" of sufficiency: there are definite limits to material desire. The Catalonia-based economist Giorgos Kallis (2016, 2019), likewise, finds in Aristotle, Solon and other Greek thinkers an ethics of *sufficiency*, *moderation* and *self-limitation*—the restriction of human desires to "the natural amount." Solon, for example, took steps to prevent "excessive accumulation." In democratic Athens, Kallis remarks, the "dominant culture was one of limits." We can learn from it in our cause of constructing a degrowth alternative to the existing order, the culture of which is "saturated with the idea of limitless accumulation" (Kallis 2019).

A final mention in this introductory survey should go to those who have found in the agronomic treatises of Antiquity evidence of commitments to frugality and self-sufficiency (for example in the work of the elder Cato), and to agroecological methods that, in their judicious interleaving of woodland, cultivated fields and animal pasture, offer a salutary contrast to modern agriculture's voracious thirst for chemical and energy inputs. In the small-scale cultivation advocated by such writers as Xenophon and Columella, some have discerned "a model of sustainability" (Hughes 2014, p. 128).

Of the arguments just reviewed, one element is uncontroversial: the ancients did live more sustainably than their counterparts today. These were low-tech non-capitalist societies, and if sustainability denotes merely the ability to use resources carefully, legions of Greek and Roman farmers and peasants deserve the badge. Knowledge of geographical and social factors that condition soil fertility, and how to mitigate soil exhaustion, was widespread. Some scholars, notably Aristotle, were aware of localised climate change, and his student, Theophrastus—hailed by some as the

Father of Ecology—recognised the role of human activity in causing it (Glacken 1967; Hughes 1975). Poets such as Horace displayed a keen sensitivity to the natural environment and spoke up for rural life. In popular idioms, the knowledge that nature forms the basis of human livelihood was woven into myths, and fertility rites. A classic is the legend of King Midas, whose prayer turned his surroundings into gold, almost leading to his death by starvation. Equally memorable is the legend of Erysichthon (literally, "earth-tearer"), a king of Thessaly. Wishing to construct a banqueting hall, he ordered his retainers to raze a sacred grove, known to be a favourite nook of Demeter, goddess of the earth's fertility. In revenge, Demeter inflicted upon Erysichthon an insatiable hunger. As much as he ate, he would always crave more. He chewed his way through his and his family's animals—the livestock, the war horse, even the cat—until ultimately he began to devour his own flesh (Adams 2012; Dale 2016).

That said, the cultivators of Antiquity were not as limit-conscious as the above authors make out (nor anywhere near as feminist as in Shiva's account).[1] Greek subsistence farmers were perfectly capable of turning a blind eye to the ecological consequences of their activity. Hunting decimated the population of larger animals, such as lions and leopards, while the domestication of sheep and goats ensured that when woodland was cleared for firewood, charcoal or timber, reforestation would be inhibited by grazing (Hughes 1975). As Greece's rich carpet of sclerophyllous woodland retreated before the axe, and double cropping sapped the fertility of the soil, the landscape faded from evergreen to a scrubby tan. The hills and mountains that had in the distant past been swathed in forest, Plato lamented in *Critias* (111a), now provide "only enough food for bees." Once diverse and bountiful, the landscape had come to resemble "the skeleton of a sick body now that the rich soft soil has been eroded, and only the thin body of the land is left."

[1] On women, Shiva's claim may apply to earlier matrilocal communities but it bears little resemblance to ancient Greece. "Oikos" is more meaningfully translated as the authoritarian (or patriarchal) household; its "economics" centrally involved the management of the patriarch's wife (Foucault 1990). In democratic Athens women may have achieved greater autonomy than women in, say, Victorian Britain, but any power they had was confined within the oikos. In public life, they were denied meaningful participation, their invisibility symbolised in the wearing of the veil (Nafissi 2004; Graeber 2011).

Evidently, Plato recognised that natural *and human* processes could occasion landscape deterioration, but did he—or Aristotle et al.—connect this environmental awareness to an ethical prescription of frugality and sufficiency? The short answer is: they did not. A longer answer, which occupies the next section, reveals that the discourse of sufficiency arose as part of a complex reaction by Greek aristocrats in the face of challenges posed by class struggle and monetisation. In the final part of the essay, I shall argue that if one looks to the ulterior purposes of ancient discourses of sufficiency (and self-sufficiency, frugality, etc.), one finds a concern for aristocratic reinvention geared to justifying structures of exploitation and imperial expansion.

2 Sufficiency and Its Ulterior Motives

Without doubt, the dominant ideologies in ancient Greece exhibited a scepticism towards "limitlessness," but what exactly was the nature of the alternative, "sufficiency"? We could do worse than begin with Aristotle. He was indeed, as Scott Cato depicts him, an advocate of the self-sufficiency of *polis* and *oikos*. In either arena, mastering the arts of goods acquisition requires a differentiation between natural and unnatural forms of exchange. When he refers to something as "natural," he means that it has within it "a principle of growth, organization, and movement" (Collingwood 1960, pp. 81–82). This is not growth in the modern sense. Rather, the goal towards which natural economic behaviour moves is the virtuous management of *oikos* and *polis*. This requires the acquisition of property (including slaves) and the provision of a supply of useful goods sufficient to afford the patriarch his freedom and comfort. Aristotle contrasts this virtuous economics with the unhealthy kind: geared to the satiating of desire, tending to excess and stimulated by coinage and the notion that the essence of wealth is a quantity of money. Up to a point, he allows, trade and acquisition (*chrêmatistikê*) serve the natural goal of economic self-sufficiency, but to treat them as ends in themselves is unacceptable. There is something about money that invites unnatural behaviour such as huckstering and accumulation for its own sake. Marcel Hénaff's exegesis of Aristotle is helpful here. Aristotle, and Plato too, viewed with concern the capacity of monetary tokens "to become autonomous and to start operating on their own without a referent, generating wealth that is unrelated to any product and thus unnatural" (Hénaff 2010, p. 79). Money, in this view, can legitimately function as a link in the economic process, an obedient handmaiden to trade, but it has another potential, for, unlike other commodities, it is "not a good as such but its expression," it only *translates* wealth (Hénaff 2010, p. 84). Whereas a good can have two functions, in consumption or exchange, money only has one, and this permits it to become detached from landed property and thus to lose its natural character. When money comes to be used to beget money, as an end in itself, use-values are shorn of their natural purpose and rendered mere "repositories of monetary value" (Skidelsky and Skidelsky 2012, p. 75). This perversion of means into ends and ends into means reaches its climax with the most unnatural kind of acquisition, usury—the creation of wealth untethered to any natural basis in productive-transformative activity.

Among Aristotle and other philosophers of Antiquity, scepticism towards money was turned to a novel and highly influential end: a self-interested defence of sufficiency. The context—and here I begin to depart from the left-Rousseauian readings—was the conjoined threat that money *and democracy* posed to traditional hierarchies, and to the oligarchy in particular. To simplify only a little, when the aristocracy found its world-view destabilised by money and democracy, political philosophy was born. Aristotle, Plato and Xenophon were seeking to reinvent the interests and identity of the nobility, without falling in with the reactionaries who wished society to be governed again by ancestral custom. The problem they faced—and here's the knot that connects money and democracy—is that the democrats were lambasting the aristocracy for their greed. This could not be rejected out of hand, for the evidence was legion. Instead, as Ryan Balot has shown (2001), they adopted a different tack.

In *The Republic*, Plato (1997) draws a correlation between the sober and moderate rule over one's inner passions and the male elite's sober and moderate rule over women and the lower orders. He distinguishes three parts of the soul: the *rational*, associated with philosophers and their wisdom, the *spirited*, associated with warriors and their qualities of pride and courage and the *appetite*, associated with producers, enslaved as they are to bodily passions and desirous of luxuries and the money to pay for them. Just as the virtuous individual is governed by the soul's rational part, with support from the spirited part while keeping the appetite under strict control, society should be governed by philosophers and warriors, who share a propensity to temperance, while the hedonistic, chaotic producers must be kept under a strict thumb. This thesis reverberated down the centuries, giving elitists a philosophical self-justification, one that comes dripping with smug self-regard and crude prejudice.

Superficially, Plato's critique of avarice speaks to radical environmentalists. But at heart, it is a conservative critique of the levelling instincts and lack of self-control of the demos. Against the democrats' accusation that the aristocrats are greedy, Plato replies that they are the very repository of reason. He justifies oligarchic domination by depicting the model aristocrat as a paragon of self-discipline. In essence: "The wellspring of greed isn't our *oikos*, it's your *demos*."

Xenophon's perspective was much like Plato's, if more pragmatic in tenor. In *Oeconomicus*, he portrays the exemplary gentleman as one who is capable of self-discipline, a trait that enables and justifies his command over others. One such is Socrates, portrayed as the model of aristocratic virtue through self-restraint. His ability to live within his means is commended as "the art of creating a surplus" (Xenophon 2008; Bosman 2015). Another is the nobleman Iskhomakhos, portrayed by Xenophon as fulfilling his virtue by directing his *oikos* to the goal of sufficiency. In William Higgins' paraphrase (1977, pp. 28–30), Iskhomakhos has "no use for stewards who are greedy," wishing them to be driven instead by a desire for his praise. In identifying the elite with values of prudence and self-restraint, Xenophon was elaborating what Balot (2001) describes as a paradigm that "recreated the aristocratic self-image in the face of the newly energized democratic ideology. ... Xenophon develops a counter-image of the morally virtuous aristocratic leader [who] deserves political prominence because of his appropriate relationship to wealth."

That democracy presented a threat to aristocratic privilege is obvious, but monetisation posed a subtler challenge. The line that defined the social world of the aristocrat was drawn between those who by virtue of their command over the labour of others were free to lead a civilised life and those who had to work to maintain themselves. The mark of a free man, as Aristotle saw it, is "not to live for the benefit of another" (Seaford, n.d., p. 161). This was captured in the notion of self-sufficiency (*autarkeia*). It combines references to the adequate supply of goods and services ("having plenty") and to an ethic of self-discipline ("managing with enough") alongside connotations of aristocratic independence (authentically free men do not depend on others for their livelihood). Self-sufficiency, in short, represented an idealised image of the aristocrat's social position (Sallares 1991, p. 298). It rested on the occlusion and naturalisation of the slavery on which the aristocrats were in reality dependent, mocking their pretensions to self-sufficiency (McNally 2020). Indeed, it positively *required* slavery, and also, indeed, the silver–slavery–imperialism nexus that represents the very antithesis of sufficiency as the term is used today. Aristotle justified the wars fought to capture slaves as part of the natural arts of wealth acquisition. He conceived of the slave as a natural extension of the master's body, the slave's labour as therefore belonging to the master, and the hunting of human beings as a natural extension of the hunting of wild beasts (Aristotle, in Ophuls 2011, p. 171). To Xenophon, the slave-mined silver from Laureion appeared as the elixir of an eternal growth in the wealth of the *polis*. As such, his goal of a self-sufficient *polis* was umbilically connected to his fetishistic view of silver: a belief in the infinitude of public wealth that silver mining was enabling (Dale 2021). But as monetisation progressed, the ideology of self-sufficiency became more problematic. The possession of money enhanced the illusion of self-sufficiency, in that it carries a power over the labour of others; the secret to the desire for money is that it promises independence (Seaford, n.d.). For Athenian elites, in addition, control of the regional money supply, enabled by the silver mines, enhanced their imperial command and thereby, in one sense at least, the self-sufficiency of their city-state. At the same time, however, money threatened to bring independence to the lower classes too—*quelle horreur!*—in particular when services to the *polis* in democratic Athens began to be paid in wage form.

Might this contradiction account for the ambivalence with which Athenian philosophers, including Plato and Aristotle, viewed money? It was dissolving the *oikos* economy that had served their class long and well. It tended to support the constitution of citizens as autonomous agents; its true home was in the *agora* (not the *oikos*) and it democratised desire, inasmuch as all classes of the citizenry were coveting the same stuff. For the aristocrats, this was undermining deference and respect for established boundaries. For all that, however, they were well placed to bend monetary relations to their interests. They howled that monetisation had thrown traditional hierarchies into disarray, and yet, as Graeber quips (2011, p. 189), what really bothered them about money "was simply that they wanted it so much."

3 Rome: Empire Without Limit

At first glance, Rome would seem to provide less inspiration for "Rousseauian" eco-criticism than Athens. Its imperial economy could be described as Classical Athens writ large: the industrial concerns and latifundia were generally bigger and the capitalists wealthier, and Rome's imperial reach was beyond compare. It experienced greater levels of commercialisation and of market-oriented agriculture. That sector, and mining too, relied heavily on slave labour, which provided the basis of senatorial wealth (Scheidel 2012). Particularly during the Republic, a continuous supply of slaves was provided by wars. These also scythed through the ranks of the *assidui*, whom the wars conscripted, dislocated and spat out indebted in a process of social polarisation that further expanded the sway of large landowners and their demand for new slaves, and onwards in a self-perpetuating cycle. In Peter Brown's description (2012, p. 15), the emperors, senators and the rich "drove a primitive system of taxation and markets to its limits, in order to perform, each year, the magic by which mere natural produce—the resources of a predominantly agrarian society—reached them as revenue, in the form of golden *solidi*." Underpinning the process, as the Empire's fundamental "exploitative force" (Wickham 1994, p. 27; Bang 2012a, p. 297), was the state, which provided two gigantic pools of demand: Rome and the army. Centred on Rome, the province of Italia came to enjoy living standards comparable to those of early modern Western Europe, thanks to virtuous circles of an increasing money supply (from war plunder and credit expansion), infrastructure spending, trade growth, urbanisation and immigration.[2]

At the cutting edge of the imperial machine, the immediate goal was capturing territory and slaves. Like Athens, Rome's wars were colossal slave raids, uprooting populations and transporting them to where they would labour for the enrichment of aristocrats. Minerals mattered too, especially bullion. Cicero (in Bang 2007) was only half-joking when he

[2] Particularly in the last few centuries BCE, population grew rapidly across much of the empire, accompanied by a rising use of building materials, of pottery, and metals such as iron and bronze, and improved diets and literacy. But annual per capita growth in Italia, possibly as high as 0.5% between 150 and 50 BCE, was the exception. For the Western Empire as a whole the figure was far lower, at perhaps 0.1% per annum between 200 BCE and CE 100 (Jongman 2014; Kay 2014; Erdkamp 2005; Morris et al. 2007; Bateman 2012).

advised his protégé Trebatius, who had joined Caesar in crossing the Oceanus Britannicus, "I hear there is not an ounce of gold or silver in Britain. If that is true, my advice is to lay hold of a chariot and hurry back to us at full speed!" Particularly during the Republic, Rome was a project built around loot. In the words of Cato (in Kay 2014, p. 28), the "desire to plunder," to possess the victims' wealth, coursed through the veins of "not a few of the most distinguished men." He was referring to the conquest of one small island, but the sentiment applied widely.

Plunder, for the imperial elite, served geopolitical and political-economic goals. The first was well captured by Polybius, the Hellenistic-era Greek historian, in his claim that Rome's armies adhered to "a certain logic in appropriating all the gold and silver for themselves; for it was impossible for them to aim at world domination unless they deprived other peoples of such resources and acquired them for themselves" (in Kay 2014, p. 27). The second was the financing of the war machine itself, as well as infrastructure and monuments, and rewarding the oligarchs, senators and legionaries, plus *panem et circenses* for the plebs, reflecting the idea that the broader citizenry should have a modest share in the spoils of war (Kay 2014).

Unsurprisingly in view of the Empire's wealth gap and relentless expansion, one cannot miss in Roman literature an awareness of the transgressing of natural limits and the snapping of boundaries. The poet Juvenal (1991, p. 34), in a satire on patron–client relations, depicts a banquet for which the mullet had to be brought in from afar because "our local waters are all fished out and exhausted," the "market's non-stop trawling" driven by "rampaging gluttony" having scoured the seas. In justifying imperial conquest, its poets invoke idioms of providence and divine mission, but also infinite expansion. We see this in Virgil's *Aeneid* (29–19 BCE, p. 13), where Jupiter declares that Rome will know "no bounds, either in space or time, ... unlimited power," prophesies the reign of Julius Caesar "whose empire shall reach to the ocean's limits," and instructs Aeneas, Rome's mythological founder, to establish an "empire without limit" (*imperium sine fine*). In Seneca's *Medea* (c. 50 CE), we find an apprehensive astonishment at imperial globalisation and the breaching of long-established limits:

> Now, in our time, the deep has ceased resistance and submits utterly to law; [...] any little craft now wanders at will upon the deep. All bounds have been removed, cities have set their walls in new lands, and the world, now

> passable throughout, has left nothing where it once had place: the Indian drinks of the cold Arazes, the Persians quaff the Elbe and the Rhine. There will come an age in the far-off years when Ocean shall unloose the bonds of things, when the whole broad earth shall be revealed, when Tethys shall disclose new worlds and Thule not be the limit of the lands.

Seneca's fable likens the thrust of technology and trade in the rise of Rome to a "trespassing over the limits posited by the gods" (Tonnetti 2016). It is Promethean, an act that humiliates the seas and destroys traditional cosmic arrangements in the course of subjugating the world to human artifice and the Roman elite's limitless quest for power.

Attitudes to the natural world manifested these norms, too. Wild nature was feared, its vanquishing was enacted symbolically in the ritual massacres of exotic beasts in public spectacles, on a scale that endangered entire species (Thommen 2012, p. 95). Roman generals and landowners were keen deforesters, in the interests of war and farmland expansion (Hughes 1975). Forest clearance was upheld as the basis of civilisation, reflected in the heroic status of such figures as Hercules, who "brought light into the gloomy forests and made the world safe for human beings" (Bergthaler 2016). The natural environment, Romans held, exists for their benefit and its resources were assumed to be continually replenishable. Although deforestation was known to lead to flooding, and resource depletion in timber, mining and farmland was periodically observed, these were "hardly perceived as alarm signals," and could be offset through annexing more territory and opening new lands for the plough (Thommen 2012, p. 13; Rimas and Fraser 2010, p. 60). Criticisms could occasionally be heard of ecological harms such as clear-cutting of forests, soil exhaustion, erosion, overgrazing and eradication of fauna, but rarely did this provoke conservation measures (Thommen 2012, p. 77). While ever more mineshafts were dug in search of metals and treasure, sanctified by the faith that nature exists for human ends, poets such as Ovid and philosophers such as Seneca—the "conscience of the Empire" (Wilson 2014, p. 229)—and the elder Pliny counselled against wanton intrusions into Mother Earth's womb (Merchant 1980). Yet if a modern sense of sustainability appears to "shimmer through" such protests, observes Thommen (2012, pp. 77, 122), it does so within a very different context: the analysis may have been "that humankind is challenging nature," but the conclusion is essentially moral, an injunction to moderate the desire for luxury, and not "an environmental call aimed at saving nature." Insofar as conservation mattered at all, it was the husbanding of resources in the pursuit of future economic gain.

4 Nemesis and Aristocratic Self-Fashioning

Famously, the commanding ethos of the Roman nobility was competitiveness and greed, manifested in ostentatious displays of wealth and the jealous amassing of property. For the wider citizenry, ambition to raise one's status, notably through gaining honours attendant on imperial service, was a must (Brown 2012, p. 25). All this hubris was self-evidently compatible with economic and military expansion, but it provoked its nemesis. Overaccumulation, as de Ste Croix explains (1981, p. 497), was a critical factor in Rome's fall: "There was nothing to restrain the greed and ambition of the rich, except in so far as the emperor himself might feel it necessary to put a curb on certain excesses in order to prevent a general or local collapse, or simply in order that the population of the empire [...] might be prosperous enough to be able to pay their taxes promptly." The concentration of power in the hands of a peculiarly exploitative and rapacious aristocracy drained the wider economy and ultimately proved to be the Empire's undoing (Croix 1981, pp. 497–502).[3]

The counterweight to aristocratic greed was class struggle—the subject of Ste Croix's book. In the ancient civilisations, the accumulation of excessive private fortunes was widely distrusted. It was, observes Michael Hudson (2001, p. 8), "perceived to sow the seeds for economic polarization, and hence social discord and decay. Down through the epoch of Roman Stoicism and into Christianity, wealth was seen to make its possessors drunk with arrogance, addicting them to seeking riches without limit in predatory ways"—and was punished by justice goddesses such as Nemesis. Within Roman elite circles, the critique of greed took religious, ethical and aesthetic forms. The governing model was "dietary" (Foucault 1990; Skidelsky and Skidelsky 2012). Much as our bodies are trained to stop eating when sated, citizens were enjoined to stop accumulating once they had piled up enough. In literary works, admonitions

[3] Hughes (1975, p. 130; 1994) has proposed that climate change and other environmental problems were factors in the fall of Rome. This thesis has recently been elaborated by Kyle Harper (2017), provoking a storm of debate that I cannot enter into here.

against elite rapaciousness and exhortations to moderation were commonplace. While ever more gold found its way into aristocrats' coffers and around noblewomens' wrists, poets such as Juvenal and Virgil inveighed against *auri sacra fames*, the idolatrous greed for gold, while Tibullus and Horace extolled *parsimonia* (frugality), and recommended a life devoted to "humility, desire for a modest fortune, quiet, physical activity, and simplicity" (Glacken 1967).

We find, then, in the Roman Empire an imperial state geared to territorial expansion, a class of landowners striving to accumulate property and certain periods (notably the Principate) that experienced productivity growth that was rapid by the standards of the era. However, peasant production remained geared largely to subsistence needs and their participation in markets, although potentially profitable, was limited—except in the vicinity of urban centres (Bang 2007). The bulk of exchanged goods circulated locally, industrial investment remained low, elite involvement in manufacturing was limited and no significant industrial bourgeoisie emerged (Anderson 1985; Bang 2012b; Garnsey 1987; Maddison 2007; Phelps 2013).

These economic structures shaped economic ethics. In the absence of an economic "system" analytically distinct from the imperial state, the Romans, like the other ancients, did not develop a meaningful concept of "the economy" as such, nor a systematic understanding of economics either as abstract theory or as a realm independent of politics (Vivenza 2012, p. 25; Finley 1973; Kay 2014). The prevailing value system was firmly that of the landed aristocracy. Roman law was permissive of trade and moneymaking but landowning was seen as the most secure and honourable occupation, and commerce was viewed askance, as a potential threat to the aristocratic way of life (Garnsey 1987). In line with the Greek philosophies discussed above, Cato (1998) disparages trade and usury and urges landowners to aspire to self-sufficiency. Cicero (44 BCE), likewise, hails agriculture as the best—the most productive and most pleasant—means of "acquiring gain," and disparages commerce as "vulgar" (unless practised without fraud and on a large scale, in which case "it is not so very discreditable" and may even be regarded with esteem, so long as the merchant, when "contented with his profits, ... betakes himself from the port itself to an estate in the country"). "Do we never find in antiquity an inquiry into which form of landed property, etc. is the most productive, creates the greatest wealth?," asks Marx (1857). Cato, he cites in response, may have investigated "which manner of cultivating a field brings the greatest rewards" and Brutus may have lent his money at high rates of interest but the prevailing question was "which mode of property creates the best citizens," and to this Rome knew a one-word answer: landowning. With this backdrop in mind, let us take another look at those agricultural authors reputed to champion principles of frugality and self-sufficiency (Cato) and sustainability (Columella).

In Cato's case, self-sufficiency and frugality cannot be divorced from his underlying goal: the justification of militarism, enslavement and aristocracy. A farmer and army commander, Cato's *De Agricultura* treats agriculture as a fulcrum of warfare: it provides the best training ground for citizen-soldiers, and upon their return from battle, farmland is their reward. *De Agricultura*, in Brendon Reay's insightful reading (2005), brings farming and tradition together in a project of "aristocratic self-fashioning." It portrays landowning as the natural home of tradition, and the landowning class as the bearers of tradition, the living link between past glories and contemporary power. But how, Reay asks, could the traditional ideal of "the farmer-statesman individually cultivating his small plot" be squared with the "contemporary aristocratic agricultural situation, characterized by absentee ownership of sometimes multiple enterprises worked primarily by slave labor?" For Cato this was not a problem and could not even be recognised as such, for he subscribed to the classical aristocratic ethos on slaves and hired hands: they are mere prostheses of their master, their work is his. Thus the absentee landowner could regard himself as the true and proper heir of the frugal farmer-statesman of yore.

De Agricultura was aimed above all at soldiers returning from the Second Punic War who had acquired the means to purchase farmland and slaves but not yet sufficient knowledge of the agronomic arts. The recommendation, in Jane Draycott's paraphrase (2019), was to achieve "economic self-sufficiency and investment potential, with as much as possible being produced on the estate." The degree to which landowners invested in improvement in order to maximise financial gain has been the subject of debate. Some economic historians suppose profit maximisation to have been uppermost in their minds; others, that the objective was to realise "a steady income from their estates with minimal intervention, and that this was best done by leasing to tenants" (Kay 2014, p. 151). Certainly, *De Agricultura* pays detailed attention to profit-loss calculations, for example with regard to slave ownership. Depicting farming as an enterprise fitted above all for absentee noble landowners who seek to maximise returns, it reveals a high level of agronomic knowledge and financial nous, notably in meticulous calculations of the costs and benefits of purchasing an olive crusher, in its advice that field hands receive a higher calorific intake during the arduous summer months and in its notorious recommendation that "old, depreciated slaves be sold in order to off-load the costs of maintenance" (Saller 2008; Kay 2014, pp. 29, 151). Cato himself became more interested in moneymaking in his later years. According to Plutarch (in Kay 2014, p. 230), he came to see "agriculture as more entertaining than profitable, and invested his capital in business that was safe and sure. He bought pools, hot springs, places given over to fullers, pitch-works, and land with natural pasturage and forests, all of which brought him much money."

Essentially the same value system continued into the Principate. The paeans to landowning soldier-farmers that we find in such treatises as Columella's *De Re Rustica* (1941, pp. 5, 9) are cut from cloth much like Cato's. Columella, who had taken to farming after a military career, expresses suspicion towards urban life, with its luxury and excess. In his valuation of a rustic existence, of "voluntary simplicity," "sufficiency" and other Stoic virtues, he presented "a conception of life that is able to separate *life quality* from *living standards*" (Keßler and Ott 2016, p. 209). And yet these commitments are shot through with ambivalence. As Keßler and Ott remark (2016, p. 210), with his economist's hat on, he "advertises the beneficial effect of selling certain products to indulge urban luxury." He enjoins estate managers to apply "measure and proportion" to their dealings, and while they may legitimately cast admiring eyes at large farms, they should till only "a small one." This may appear *prima facie* as a Roman "small is beautiful," but in fact the goal is profitability. One should, Columella goes on, avoid succumbing to the "unrestrained passion for ownership," for if one does, one's cultivable lands will be left to decay and *lie unprofitably fallow*. Therefore, "as in all matters, so too in the acquiring of land, moderation shall be exercised" (Columella 1941, pp. 47–57). Estate managers, he goes on, should invest to improve their vineyards, but they should also go beyond that, boldly and systematically experimenting, in order that the soil become fitted to its best use and yield the greatest profit (Columella 1941, p. 55; Rostovtzeff 1926, p. 93). Throughout *De Re Rustica*, land's telos is to be of use to the farmer; the argument is emphatically economic, with no consideration for ecological thresholds and constraints (Keßler and Ott 2016, p. 210).

5 The Slippery Gold of Seneca

Columella (1941, p. 255) reserves the highest praise ("outstanding genius and erudition") for his friend Seneca, the Stoic philosopher who is lionised today by advocates of consumption ethics of sufficiency. Seneca, as interpreted by Kai Whiting and Leonidas Konstantakos (2019), "explores the link between greed (a Stoic vice) and environmental deterioration." As evidence, they cite his address to noblemen whose self-indulgence and greed had spawned orgies of villa construction:

> I ask you: how long will this go on? Every lake is overhung with your roofs! Every river is bordered by your buildings! Wherever one finds gushing streams of hot water, new pleasure houses will be started. Wherever a shore curves into a bay, you will instantly lay down foundations. [...] Yet when you have done so much enormous building, you still have only one body apiece, and that a puny one. What good are numerous bedrooms? You can only lie in one of them.

As Lukas Thommen has pointed out (2012, p. 3), Seneca here is berating his audience not for environmental crimes but for their "material greed and addiction to splendour." Seneca did diagnose excessive luxury as a crime against "nature," but the moral prescriptions that follow were not that one should protect nature, but, rather, one should fit in "sensibly with the existing world order" (Thommen 2012, p. 78).

In their blistering denunciation of avarice, Seneca's lines on lakeside villas are characteristic. He regarded avarice as natural, or at least ubiquitous, but reprehensible. Just as "children are greedy for knuckle-bones, nuts and coppers," he sighed, "grown-ups are greedy for gold, silver and cities" (in Wilson 2014, p. 31). In one delightful fable that purports to explain the rarity and the dangers of bullion, he tells of Earth's having revealed "everything that was likely to be of use" to the human race, while making sure to hide the gold and silver. These she buried "deep, and weighted them down with all her mass, regarding them as harmful substances, destined to be a curse to the nations if brought forth into the light." Yet, our Stoic sage continues, no less dangerous than bullion are securities and bonds, usury and accounting, those "empty phantoms of ownership" that attract "unnatural forms of human greed." Unlike gold and silver, such evils originate in human minds; they are "the mere dreams of empty Avarice." For true fulfilment one must turn not to wealth but to wisdom and morality. Wretched is he who takes delight in the expanse of his estate, "in his vast tracts of land that need to be tilled by men in chains, in huge herds and flocks that need whole provinces and kingdoms to provide them with pasture, and in private palaces that cover more ground than great cities!" For after he, "puffed up with pride," has tallied his possessions he will at once turn attention to "what he still covets," and in so doing is again confronted with his still-insufferable poverty (Seneca, ca. 60). This hankering for wealth, and the desires that underpin it, should be curbed in favour of a life of sufficiency, even frugality, for only thus will we be freed from the vagaries of fortune, and able to live in authentic self-dependence (Wedeck 1955, p. 540).

These lines from *On Benefits* are often taken to be rebukes of social polarisation. In fact they are the very opposite. They offer the wealthy a subtle ethical defence of their fortunes. Notice, first, that Seneca's concern lies not with the "men in chains" sweating in the fields but with the slave-owner whose covetousness prevents him from enjoying the serenity that an ethical approach would bring.[4] As his biographer Emily Wilson explains, he had no interest in wealth redistribution or in challenging the conditions of poverty or slavery, and he "barely touches on whether the acquisition and possession and use of wealth by an individual represents and causes significant social injustice" (Wilson 2014, pp. 139, 140, 189). His central message, and that of Stoicism generally, is that living virtuously depends on the inner life, with mastery of the passions and the achievement of inner calm, requiring an indifference to the social relationships in which one is situated—including hierarchies of wealth and status. Individuals should not, Seneca maintains, think of wealth or social polarisation as a social-structural matter, but as a source of challenges and lures that the wealthy individual must rise above (Wilson 2015). He is concerned above all to enable the rich to cope psychologically with their wealth, to alleviate their thraldom to consumer*ism*, regardless of how much they *consume*. The goal is not simplicity but contentment. In his words: "He is a great man who uses earthenware dishes as if they were silver; but he is equally great who uses silver as if it were earthenware" (Wilson 2014, p. 186).

When reading Seneca one cannot but wonder whether this preacher of frugality, of self-restraint and moderation of the appetites, this critic inveighing against property as humanity's greatest affliction, can be the

[4] Seneca's critical defence of slavery is in Seneca (n.d. a).

same Seneca who amassed a fabulous private fortune, the heir, the major slave-owner, the recipient of a small fortune from Nero (as payment for his flattery)—the very same, indeed, who was able to order five hundred tables made of the finest citrus wood with legs of ivory, "all exactly identical" (and this in an age of hand manufacture in which identical items of such sophistication were fiendishly difficult to make) (Wilson 2014, p. 128). The charge of hypocrisy, however, does not take us far. More interesting is that Seneca's ethical system serves precisely to *protect* aristocrats from the smell of hypocrisy. In this worldview, "frugal plutocrat" is no oxymoron, for frugality is "an inner disposition, not an outward affectation" (Gildenhard 2020, p. 319).

Seneca was writing during the Principate, a time in which unbridled oligarchical power was heightening the pressure on elites to justify their status and conspicuous consumption. One means of doing so was civic euergetism: performing *euergesiai* (good deeds) for one's city (Brown 2012). The practice, rooted in Greek elite gift-giving and liturgy, blended with traditions of patronage. In the Greek ethical code, economic desert was closely linked to moral excellence. Only those who possessed the appropriate moral virtues were fitted to possess wealth and power (Zuiderhoek 2009). Euergetism was a means of demonstrating one's virtue through public largesse, and hence of justifying one's wealth. These were "politico-ritualistic acts" of public munificence that served "to underwrite the ideal of the essential unity" of the citizen community and to legitimate the oligarchic concentration of wealth and power (Zuiderhoek 2009). Euergetism served a practical purpose; it cemented the personal ties that held Roman society together. But its role was powerfully ideological too, in that it established the fiction that the cities of the Empire were, in Arjan Zuiderhoek's description (2009), "organic, closely-knit and cohesive citizen communities," rather than "the polarised societies strongly divided by enormous disparities of wealth and political power that they were in reality."

Under Octavian, when Seneca was young, Rome was awash with commercialism and mired in corruption. Its nobility struggled to present an upstanding self-image. Octavian reinvented his persona as Augustus: emperor in status, but frugal, benevolent and civic-minded republican of soul. As a counterfoil he cultivated a mythical image of his arch-rival, Cleopatra, as one of oriental excess, the decadent drunken whore-queen who pisses in a gold chamber pot. All this, remarks William Adams (2012, pp. 64–65), was "a brilliant move to exorcise the taints of corruption and dissolution that were associated with the growing spread of private profligacy." Stoicism, with its sensibility of moderation, and its ethics of "family values" and civic order, gained a considerable following. Its ethic of frugality suited the imperial throne as it attempted to tamp down inter-oligarchical conflict. It spoke to the inner conflicts of elites who had grown increasingly materialistic and consumerist as the empire expanded (Wilson 2014, p. 199).

This formed the backdrop to *On Benefits*, Seneca's essay on the importance of gift-giving as the moral core of human society. Giving, he argues (Seneca, ca. 60 CE), is "a social act" that wins the goodwill of others, establishing ties that bind the givers and recipients. It gestures towards genuine community, in which hierarchical relationships recede into the background, supplanted by relationships of honesty and dignity in which neither party seeks to profit at the expense of the other (Wilson 2014, p. 147).

At one level, by affirming the code of gift-exchange within the upper orders, a code that was continuous with tradition, with the Republic, Seneca was implicitly giving his blessing to the Principate.[5] Yet he was also doing something quite new. He identifies a quality in the act of giving that is similar to *selling*. The latter, he explains, means "alienation, the transferring of one's property and one's right in it to another." But the same applies to giving. It entails "the relinquishment of something, the surrendering of something that you have held to the possession of another" (Seneca, ca. 60). This is a redefinition of gift-giving as something undertaken freely *by the individual*, governed by ethical intention alone. It is a unilateral act, unlike the traditional, ritual form—and in this, Seneca is reminding his readers that ceremonial gift-exchange had become a distant memory (Hénaff 2010, pp. 259–260). However, if it is in a formal sense akin to selling, the acts of merchants, giving represents something profoundly different. Selling is a niggardly and miserly practice, unlike authentic, aristocratic gift-giving which is generous—as exemplified in the purchase of the five hundred tables mentioned above.

Building on Greek foundations, Seneca developed an ethics that brings together, as elements of the virtuous life, a communitarian concern for social solidarity (specifically, within the aristocratic caste) and an advocacy of independence and self-sufficiency—much as in Aristotle, in the sense

[5] Seneca's implicit defence of the Principate is discussed in Griffin (2013, p. 11).

of one's imaginary economic independence from others. But in Seneca the emphasis on individual moral responsibility is more pronounced. The capacity for practical, rational deliberation makes the individual free even in relation to his own character, and therefore morally responsible (Engberg-Pedersen 1990, p. 131).

What is striking is that Seneca is taking moral individualism, as it had developed for example in the thought of the Sophists and Socrates, a leap forward, in response to the more monetised environment in which he lived (Hénaff 2010). That he draws a comparison between giving and selling is symptomatic. The commercial mindset was in the ascendant and a reciprocal economy based on the services of friends, family and peers was being edged out by commercial and litigious transactions (Griffin 2013, pp. 54–61). It is in this context that his presentation of giving as a unilateral act, governed by the giver's intention alone, should be understood. His identification of the alienation of one's possessions as the common denominator of giving *and selling* bespeaks a culture in which private property norms had become entrenched and absolutised.

Seneca's ethical individualism, I am suggesting, facilitated his evasive moves on questions of wealth inequality. It allowed one to be fantastically wealthy without compromising one's ethical integrity. If ethics is about the individual achieving happiness (or sufficiency) based on the capacities that they have within themselves to change their lives (Asmis 2015), what matters is not how one's wealth contributes to the relative poverty of others but solely how one gained it and how one wears it. Seneca elaborates an ethics—in effect it's a "Just Accumulation Theory"—that, in focusing on individual behaviour and "character," can justify the foulest exploitation. Racking up a fortune is fine so long as it's acquired with a smile and a handshake, at least some commitment to frugality in one's consumption habits, a sense of sufficiency and capped with generous dollops of philanthropy. Taken together, these acts and attitudes will suffuse the soul with grace and inner peace. Little wonder that, in spite of his often desolate tone and his distaste for technology, Seneca finds favour among Silicon Valley's "self-help gurus" today (Mead 2011).

But Seneca goes further still. Albeit in a somewhat ambiguous text, he appears to suggest that avarice, excess and luxury gave birth not only to private property but also to morality itself (Straumann 2015, p. 185). This idea is nested within a narrative of social decline and progress, beginning with the original state of nature that is known (inaptly in Seneca's case) as the golden age. Before I summarise his account, I should sketch the discursive context. Not unlike us "moderns," Antiquity knew its own narratives of civilisational progress and regress. (Indeed, in this sense the ancients were "moderns.") In the progress narrative, humanity was said to have originated in a rudimentary condition of suffering and hardship, later evolving, thanks to technology, morality and politics, to a civilised state (Thommen 2012, p. 31). In the alternate narrative, humanity had originally been blessed with natural abundance. The soils were rich and required little or no agricultural toil. Despite the lack of developed technologies and conveniences, life was simple and contented (Hesiod 1988). In the version of Dicaearchus, a student of Aristotle and friend of Theophrastus, the original age was happy because people did not crave amenities the production of which would require hard graft. Since they knew "sufficiency," they had few cares and no tussles over resources (Glacken 1967, p. 140).

Distinctive in Seneca is his dialectical sublation of the progress and regress narratives. The original age, he speculates, was halcyon and communistic but also pre-moral. It preceded the onset of the cleft between nature and culture. Nature was bountiful, before greed and property intervened. Human needs in that fortune-favoured time were met by nature, "and this her gift consisted of the assured possession by each man of the common resources." It was an age of abundance, of sufficiency and therefore harmony. The soil

> yielded more than enough for peoples who refrained from despoiling one another. [...] They searched not in the lowest dregs of the earth for gold, nor yet for silver or transparent stones; and they still were merciful even to the dumb animals. [...] They had as yet no embroidered garments nor did they weave cloth of gold; gold was not yet even mined. (Seneca, n.d. b)

Then along came the gold diggers, bringing discord. Avarice "broke in upon a condition so happily ordained, and, by its eagerness to lay something away and to turn it to its own private use, made all things the property of others, and reduced itself from boundless wealth to straitened need" (Seneca, n.d. b). Together, avarice and luxury snapped the bonds "which held mortals together, and they, abandoning their communal existence, separated and turned to plunder" (Seneca, n.d. b). At this point in the account something dramatically new occurs: the war of all against all gives way to culture and law, for these were necessary to mitigate and manage vice, and to enable social progress. This is, so to speak, a "cunning of virtue," in that avarice brought not only plunder but culture and civilisation. In the state of nature, no skilled artisans had been required; "it is only the luxuries that call for labour" (Seneca, n.d. b). With labour and luxury came a flourishing of skills, the arts, culture and reason itself—and only with these could morality be knowingly and therefore meaningfully exercised. Avarice, paradoxically, was the midwife of virtue.

Seneca's philosophy of justice aligned with early Christian thought and influenced it too. This was the case in respect of his views on frugality and asceticism (which fed into the monastic injunction to voluntary poverty and "sufficiency" as necessary for spiritual devotion); his inveighing against covetousness (coupled with a refusal to lift a finger against structural causes of wealth polarisation); his justification of wealth possession through gift-giving (albeit in Christianity the envisaged recipient shifts from fellow aristocrats and townsfolk to the Church and the poor) and his concept of giving as a unilateral act (akin to the Christian principle of charity as unadulterated altruism). Whether or not Seneca's views were influenced by the Christians, they came to embrace him as one of their own.

6 Conclusion

Part of the appeal that Greek and Roman Antiquity exerts on the present stems from that era's vivid documents of the ability of humans to control their natural environment. In agricultural writings, from Xenophon to Virgil and beyond, we read of a growing recognition of humanity's ability to shape the natural order. With only few exceptions, however, the goal was straightforwardly "economic": to increase agricultural surplus (cf. Dale 2021). While energy use and resource throughput were incomparably lower than in today's world, these were not "green" economies in any meaningful way; sustainability is incompatible with modes of production based on slavery, war and the dominium of landed aristocrats. Attitudes to the natural realm varied widely, and included the worship of deities manifesting natural phenomena, the sparing of some animals by hunters, and, in some localities, forest protection by professional wardens, but on the whole, a social order based on the exploitation of humans and the expansion of empire encouraged an instrumental attitude that perceived nature as a resource for appropriation. Nor did a meaningful concept of sustainability exist. As Thommen (2012, p. 8) has argued, the ancients had not addressed questions of "the planned management and distribution of resources," and the question, too, of equal access to goods and social justice was "completely beyond the pale."

The sustainability discourse is thoroughly modern, and so too are strategies to achieve it via "sustainable consumption"—a lifestyle of "treading lightly" and "sufficiency," with the aim of reducing society's environmental footprint. No sufficiency-sustainability link was posited by the ancients. We may glimpse it in some nineteenth-century authors, such as Thoreau,[6] but it did not come into its own until the 1960s—a decade shaken by environmentalism's breakthrough as a mass social movement, by unprecedented economic growth as well as critiques of consumer culture, the birth of ecological economics and the first diagnoses of "growth fetishism" (Kidron 1966). These critiques threaded together with a variety of strands, including romantic and religious critiques of modernity (e.g. Rousseau, Illich), Marxism (Fromm, Marcuse), and sociologies of consumption (Veblen), and have bequeathed, more recently, the degrowth movement, its compass set to frugality and sufficiency.

One cannot object to the appropriation by eco-radicals of ancient concepts—Aristotle's sufficiency, or Seneca's frugality. Finding inspiration in ancient ethical codes, stripping them of their context and applying them to present-day debate is innocuous. On one point, however, we may reasonably pick a nit or two. It concerns the divorcing of ethical and psychological discourse from its social integument. The standout example from Antiquity is Seneca, a thinker who diagnosed every angle of a cultural symptom, the link of moral decadence to luxury, without once addressing its cause: a mode of production that was funnelling wealth in stupendous quantities from territory and workers (enslaved and hired) to the aristocracy, including Seneca himself. Environmentalism in our day is no stranger to such an "ethic." Consider the case of Laurance Rockefeller. One of America's most celebrated environmentalists, he was dedicated to creating "a conservation ethic that is endurable, bipartisan, and rooted in a consistent sense of values" (Winks 1997). Throughout, however, he remained a major investor in the arms sector, airlines, property development and tourism—and all this before we even mention his inherited fortune from Standard Oil.

[6] Thoreau's ethic of simplicity was designed in part with nature conservation in mind, but was principally aesthetic and spiritual: to enable a contemplative, awakened sensibility.

Such parallels should not be taken too far. In Seneca's Rome, and in pre-modern empires in general, consumption patterns were taken to reflect "natural" divisions of social status. The luxuries enjoyed by the leisure classes were what set them apart (Gellner 1983). That began to change with the rise of capitalism. Georgian Britain saw the birth of a new ideology of consumption: the idea that our lust for "stuff" is what keeps the wheels of industry moving, bringing prosperity to the whole nation. There was an imperial angle, too. The consumption of "colonial goods" (tea, tobacco and sugar), more than wars and jingoism, seduced the British public into symbolic investment in the Empire.

Consumerism, I am suggesting, should be understood as an ideology organic to capitalism. Business owners are systematically obliged to suppress the consumption of their workforce and yet they must simultaneously realise surplus value—ergo, the capitalist class as a whole relies on the stimulation of new "needs," requiring the ever-increasing purchase of new commodities (Bhattacharya 2015). This was recognised long ago by the Irish socialist William Thompson. He saw that the capitalist system requires, on one hand, the production of an increasing abundance of products, and, on the other, an "abundance of consumers to use the articles produced, so that every year a continual demand should be kept up for these or similar articles" (Thompson 1824, p. vi). In the twentieth century, the age of mass consumption, capitalists' organisations turned to the active encouragement of consumerism. After the Second World War, for example, the OECD, seeking to restabilise global capitalism, advocated that the working masses be drawn through consumption choices to identify with "the economy" (Schmelzer 2016).

In one sense, the Consumer Age represented a democratic breakthrough. Whereas the rich had once monopolised the comforts and conveniences, some were becoming available to the masses. Workers who had known destitution could obtain items that had hitherto been out of reach, and which consumer society was at last making available, pushed onwards by trade-union struggle. Consumption began to be celebrated as the means for the denizens of mass society to express their individuality, freedom and identity (Simmel 2004). No longer did the boss drive to work while the workers walked or took the bus. Now they too could drive. In other respects, both social and ecological, "growth-driven consumerism," as Kate Soper calls it (2020), has been a recipe for catastrophe.

If consumerism is an ideology of capitalism, a progressive anti-consumerism, it follows, would aim to cut the fuel supply to capital accumulation, relieving the rich of their fortunes while working to ensure that those who have accumulated nothing can at last fulfil their needs (Rodríguez-Labajos et al. 2019). Understood in these terms, sufficiency would appear in a radical-democratic light: not as the ascetic conservatism of an Ophuls or as an individual moral disposition, but as a demand for an "alternative hedonism" or, in its broadest conception, as class struggle (Soper, 2020; Barker and Dale 1998).

References

Adams, W. (2012), *On Luxury: A Cautionary Tale*, Potomac Books.

Ambrose, A. (1961), *Hexameron, Paradise and Cain and Abel*, Catholic University of America Press

Anderson, P. (1985), *Passages from Antiquity to Feudalism*, Verso.

Asmis, E. (2015), 'Seneca's Originality,' in Shadi Bartsch and Alessandro Schiesaro, eds., *The Cambridge Companion to Seneca*, Cambridge University Press.

Balot, R. (2001), *Greed and Injustice in Classical Athens*, Princeton University Press.

Bang, P. (2007), 'Trade and Empire: In Search of Organizing Concepts for the Roman Economy,' *Past and Present*, May.

Bang, P. (2012a), 'Bazaars, Empires, and Roman Trade,' in Walter Scheidel, ed., *The Cambridge Companion to the Roman Economy*, Cambridge University Press.

Bang, P. (2012b), 'Predation,' in Walter Scheidel, ed., *The Cambridge Companion to the Roman Economy*, Cambridge University Press.

Barker, C. and Dale, G. (1998), 'Protest Waves in Western Europe: A Critique of "New Social Movement Theory",' *Critical Sociology*, 25 (1–2).

Bateman, V. (2012), *Markets and Growth in Early Modern Europe*, Pickering & Chatto.

Bergthaler, H. (2016), 'Response,' in C. Schliephake, ed., *Ecocriticism, Ecology, and the Cultures of Antiquity*, London: Lexington.

Bhattacharya, T. (2015), 'How Not To Skip Class: Social Reproduction of Labor and the Global Working Class,' https://viewpointmag.com/2015/10/31/how-not-to-skip-class-social-reproduction-of-labor-and-the-global-working-class/.

Bosman, P. (2015), 'Ancient Debates on *Autarkeia* and Our Global Impasse,' *Phronimon*, 16 (1), 1–16.

Brown, P. (2012), *Through the Eye of a Needle: Wealth, the Fall of Rome and the Making of Christianity in the West, 350-550AD*, Princeton University Press.
Cato, M. P. (1998 [ca. 155 BCE]), *On Farming*, Prospect Books.
Cato, M. S. (2009), *Green Economics: An Introduction to Theory, Policy and Practice*, Earthscan.
Cicero, M. T. (44 BCE), *On Moral Duties*, www.perseus.tufts.edu/hopper/text.
Collingwood, R. (1960 [1945]), *The Idea of Nature*, Oxford University Press.
Columella, L. J. M. (1941), *On Agriculture*, William Heinemann Ltd.
Croix, G. de Ste (1981), *The Class Struggle in the Ancient Greek World: From the Archaic Age to the Arab conquests*, Duckworth.
Dale, G. (2016), 'Origins and Delusions of Green Growth,' *International Socialist Review*, https://isreview.org/issue/97/origins-and-delusions-green-growth/index.html.
Dale, G. (2021), 'The Infinite Silver of Xenophon: Money and Growth in Classical Athens,' in Nicola Ialongo *et al.*, eds., *Merchants, Measures and Money: Understanding the Technologies of Early Trade in a Comparative Perspective*, Wachholtz Verlag.
Draycott, J. (2019), *Roman Domestic Medical Practice in Central Italy: From the Middle Republic to the Early Empire*, Routledge.
Engberg-Pedersen, T. (1990), 'Stoic Philosophy and the Concept of the Person,' in Christopher Gill, ed., *The Person and the Human Mind: Issues in Ancient and Modern Philosophy*, Clarendon Press.
Erdkamp, P. (2005), *The Grain Market in the Roman Empire: A Social, Political and Economic Study*, Cambridge University Press.
Finley, M. (1973), *The Ancient Economy*, Chatto & Windus.
Foucault, M. (1990), *The History of Sexuality, Vol. 2: The Use of Pleasure*, Vintage.
Garnsey, P. (1987), *The Roman Empire: Economy, Society, and Culture*, University of California Press.
Gellner, E. (1983), *Nations and Nationalism*, Cornell University Press.
Gildenhard, I. (2020), '*Frugalitas,* Or: The Invention of a Roman Virtue,' in Ingo Gildenhard and Cristiano Viglietti, eds., *Roman Frugality: Modes of Moderation from the Archaic Age to the Early Empire and Beyond*, Cambridge University Press.
Glacken, C. (1967), *Traces on the Rhodian Shore: Nature and Culture in Western Thought from Ancient Times to the End of the Eighteenth Century*, University of California Press.
Graeber, D. (2011), *Debt: The First 5,000 Years*, Melville House.
Griffin, M. (2013), *Seneca on Society: A Guide to De Beneficiis*, Oxford University Press.
Harper, K. (2017), *The Fate of Rome: Climate, Disease, and the End of an Empire*, Princeton UP.

Hénaff, M. (2010), *The Price of Truth: Gift, Money, and Philosophy*, Stanford University Press.
Hesiod (1988), *Works and Days*, Translation M. L. West, Oxford University Press.
Higgins, W. (1977), *Xenophon the Athenian: The Problem of the Individual and the Society of Polis*, State University of New York Press.
Hudson, M. (2001), 'Mesopotamia and Classical Antiquity,' in Robert Andelson, ed., *Land-Value Taxation Around the World: Studies in Economic Reform and Social Justice*, Wiley-Blackwell
Hughes, J. D. (1975), *Ecology in Ancient Civilizations*, Albuquerque: University of New Mexico Press
Hughes, J. D. (1994), *Pan's Travail: Environmental Problems of the Ancient Greeks and Romans*, Johns Hopkins University Press.
Hughes, J. D. (2014), *Environmental Problems of the Greeks and Romans: Ecology in the Ancient Mediterranean*, Johns Hopkins University Press.
Jongman, W. (2014), 'Re-Constructing the Roman Economy,' in Larry Neal and Jeffrey Williamson, eds., *The Cambridge History of Capitalism, Volume I, The Rise of Capitalism: From Ancient Origins to 1848*, Cambridge University Press.
Juvenal (1991 [c. 112 CE]), *The Satires*, translated by Niall Rudd, Oxford University Press.
Kallis, G. (2016), 'The Vocabulary of Degrowth: A Roundtable Debate,' *Ephemera: Theory and Politics in Organization*, www.ephemerajournal.org/contribution/vocabulary-degrowth-roundtable-debate-0.
Kallis, G. (2019), *Limits: Why Malthus Was Wrong and Why Environmentalists Should Care*, Stanford University Press.
Kay, P. (2014), *Rome's Economic Revolution*, Oxford University Press.
Keßler, L. and Ott, K. (2016), 'Nec provident futuro tempori, sed quasi plane in diem vivant—Sustainable Business in Columella's De Re Rustica?' in C. Schliephake, ed., *Ecocriticism, Ecology, and the Cultures of Antiquity*, London: Lexington.
Kidron, M. (1966), 'Some Trees, No Wood,' *International Socialism*, 24, www.marxists.org/archive/kidron/works/1966/xx/nowood.htm.
Maddison, A. (2007), *Contours of the World Economy, 1-2030 AD: Essays in Macro-Economic History*, Oxford University Press.
Marx, K. (1857), *Grundrisse*, www.marxists.org/archive/marx/works/1857/grundrisse/ch09.htm.
McNally, D. (2020), *Blood and Money: War, Slavery, Finance, and Empire*, Haymarket.
Mead, R. (2011), 'Better, Faster, Stronger: Silicon Valley's Self-Help Guru,' www.newyorker.com/magazine/2011/09/05/better-faster-stronger.

Merchant, C. (1980), *The Death of Nature: Women, Ecology and the Scientific Revolution*, HarperCollins.
Morris, I., Saller, R., and Scheidel, W. (2007), 'Introduction,' in Walter Scheidel, Ian Morris and Richard Saller, eds., *The Cambridge Economic History of the Greco-Roman World*, Cambridge University Press
Nafissi, M. (2004), 'Class, Embeddedness, and the Modernity of Ancient Athens,' *Comparative Studies in Society and History*, 46 (2).
Ophuls, W. (1992), *Ecology and the Politics of Scarcity Revisited: The Unraveling of the American Dream*, W. H. Freeman and Company.
Ophuls, W. (2011), *Plato's Revenge: Politics in the Age of Ecology*, Cambridge: MIT Press.
Phelps, Edmund (2013), *Mass Flourishing: How Grassroots Innovation Created Jobs, Challenge, and Change*, Princeton University Press.
Plato (1997), *Complete Works*, edited by John Cooper, Hackett Publishing.
Quilley, S. (2013), 'De-Growth Is Not a Liberal Agenda: Relocalization, and the Limits to Low Energy Cosmopolitanism,' *Environmental Values*, 22, 261–285.
Quilley, S. (2017), 'Navigating the Anthropocene: Environmental Politics and Complexity in an Era of Limits,' in Peter Victor and Brett Dolter, eds., *Handbook on Growth and Sustainability*, Edward Elgar.
Reay, B. (2005), 'Agriculture, Writing, and Cato's Aristocratic Self-Fashioning,' *Classical Antiquity*, 24 (2), 331–361.
Rimas, A. and Fraser, E. (2010), *Empires of Food: Feast, Famine, and the Rise and Fall of Civilizations*, Simon and Schuster.
Rodríguez-Labajos, B., *et al.* (2019) 'Not So Natural an Alliance? Degrowth and Environmental Justice Movements in the Global South,' *Ecological Economics*, 157, 175–184.
Rostovtzeff, M. (1926), *The Social & Economic History of the Roman Empire*, Biblo & Tannen.
Saller, R. (2008), 'Human Capital and the Growth of the Roman Economy,' Princeton/Stanford Working Papers in Classics, www.princeton.edu/~pswpc/pdfs/saller/060809.pdf.
Sallares, R. (1991), *The Ecology of the Ancient Greek World*, Cornell University Press.
Seaford, R. (n.d.), 'Reading Money: Leslie Kurke on the Politics of Meaning in Archaic Greece,' www.bu.edu/arion/files/2010/03/Seaford-Reading-Money.pdf.
Scheidel, W. (2012), 'Slavery,' in Walter Scheidel, ed., *The Cambridge Companion to the Roman Economy*, Cambridge University Press.
Schmelzer, M. (2016), *The Hegemony of Growth, The OECD and the Making of the Economic Growth Paradigm*, Cambridge University Press.

Seneca, L. (n.d. a), Moral letters to Lucilius/Letter 47, https://en.wikisource.org/wiki/Moral_letters_to_Lucilius/Letter_47.

Seneca, L. (n.d. b), Moral letters to Lucilius/Letter 90, https://en.wikisource.org/wiki/Moral_letters_to_Lucilius/Letter_90.

Seneca, L. (ca. 50 CE), *Medea*, www.theoi.com/Text/SenecaMedea.html.

Seneca, L. (ca. 60 CE), *On Benefits*, www.stoics.com/seneca_essays_book_3.html.

Shiva, V. (2013), 'How Economic Growth Has Become Anti-Life,' www.theguardian.com/commentisfree/2013/nov/01/how-economic-growth-has-become-anti-life.

Simmel, G. (2004 [1900]), *The Philosophy of Money*, edited by D. Frisby, Routledge, London.

Skidelsky, R. and Skidelsky, E. (2012), *How Much Is Enough? The Love of Money, and the Case for the Good Life*, Allen Lane.

Soper, K. (2020), *Post-Growth Living: For an Alternative Hedonism*, Verso.

Straumann, B. (2015), *Roman Law in the State of Nature*, Cambridge University Press.

Thommen, L. (2012), *An Environmental History of Ancient Greece and Rome,* Cambridge University Press.

Thompson, W. (1824), *An Inquiry into the Principles of the Distribution of Wealth Most Conducive to Human Happiness; Applied to the Newly Proposed System of Voluntary Equality of Wealth*, Longman & Co., London.

Tonnetti, A. (2016), 'Breaking the Natural Limits Given by the Sea: The Case of the Roman Empire and Its Tragedy in the Medea of Seneca,' Conference Paper, Historical Materialism, London.

UNEP (2020), Emissions Gap Report, www.unenvironment.org/emissions-gap-report-2020.

Virgil (29–19 BCE), *The Aeneid*, Doubleday.

Vivenza, G. (2012), 'Roman Economic Thought,' in Walter Scheidel, ed., *The Cambridge Companion to the Roman Economy*, Cambridge University Press

Wedeck, H. (1955), 'The Question of Seneca's Wealth,' *Latomus*, T. 14, Fasc. 4, Octobre-Decembre.

Whiting, K., *et al.* (2018), 'Sustainable Development, Wellbeing and Material Consumption: A Stoic Perspective,' *Sustainability*, 10.

Whiting, K. and Konstantakos, L. (2019), 'Stoic Theology: Revealing or Redundant?' *Religions*, 10.

Wickham, W. (1994), *Land and Power: Studies in Italian and European Social History, 400–1200*, British School at Rome.

Wilson, E. (2014), *Seneca: A Life*, Allen Lane.

Wilson, E. (2015), 'Seneca, the Fat-Cat Philosopher,' Guardian, www.theguardian.com/books/2015/mar/27/seneca-fat-cat-philosopher-emily-wilson-a-life.

Winks, R. (1997), 'Laurance S. Rockefeller: Catalyst for Conservation,' www.washingtonpost.com/wp-srv/style/longterm/books/chap1/laurancesrockefeller.htm.

Xenophon (2008 [ca. 362 BCE]), *Oeconomicus,* www.gutenberg.org/files/1173/1173-h/1173-h.htm.

Zuiderhoek, A. (2009), *The Politics of Munificence in the Roman Empire: Citizens, Elites and Benefactors in Asia Minor*, Cambridge University Press.

Linking the Developmental State to Green Economic Growth

Philip Arestis, Nikolaos Karagiannis, and Zagros Madjd-Sadjadi

1 Introduction

Twenty-eight years after the first 'Rio Summit', the world capitalist economy continues to face the twin challenge of expanding economic opportunities for an increasing global population and addressing mounting environmental pressures. This twin challenge gave birth to the contemporary concept of 'green growth', which meant to take full account of the complex consequences of greening the growth dynamics of modern economies. Green growth means ensuring economies continue to grow while simultaneously creating a sustainable pathway so that those ecological assets that human being relies upon are there well into the future. Subsequently, the notion of 'green economy' has been accepted as a key driver in reducing pollution, tackling climate change, dealing with

P. Arestis (✉)
Department of Land Economy, University of Cambridge, Cambridge, UK
e-mail: pa267@cam.ac.uk

N. Karagiannis · Z. Madjd-Sadjadi
Winston-Salem State University, Winston-Salem, NC, USA

N. Karagiannis and J. E. King (eds.), *Visions and Strategies for a Sustainable Economy*, Global Institute for Sustainable Prosperity,
https://doi.org/10.1007/978-3-031-06493-7_2

poverty and income inequality, improving public health, and pursuing a number of critical goals to improve life for this planet and its people.

Indeed, the green aspect, mainly natural capital, is the most constrained asset owed to modern economic growth patterns. But since 2008, due to the focus on the environment-economy nexus by green growth, the term has been extensively utilized as the policy response to various global crises. During the Global Financial Crisis (GFC) of 2007–2009, the United Nations Environment Programme (UNEP) "championed the idea of 'green stimulus packages' and identified specific areas where large-scale public investment could kick-start a 'green economy'" (United Nations, n.d.).

This notion inspired international institutions as well as several national governments around the world to formulate and employ significant 'green stimulus' frameworks bringing economic, social, environmental, technological, and development aspects into comprehensive policy packages to aid in their economic recovery. Subsequently, the concept of green economic growth has now entered the collective conscience of every organization from the largest nations to the smallest communities.

- In 2008, UNEP led the 'Green Economy Initiative'.
- In 2010, S. Korean President Lee Myung-bak launched the Global Green Growth Institute (GGGI) as a think tank. The 2012 'Rio+20 Summit' in Brazil would later transform it into an international treaty-based organization.
- In 2011, the International Chamber of Commerce's (ICC) Task Force on Green Economy examined what it would take for the global economy to become green and published a series of ten different conditions that would need to be met (Bacher 2015).
- In 2011, the OECD published a green growth strategy.
- In January 2012, the Global Green Growth Institute, OECD, UNEP, and World Bank formally launched the *Green Growth Knowledge Platform* (GGKP). The mission of the GGKP "is to enhance and expand efforts to identify and address major knowledge gaps in green growth theory and practice, and to help countries design and implement policies to move towards a green economy" (GGKP 2013: 4).
- In 2012, the World Bank published *Inclusive Green Growth: The Pathway to Sustainable Development*.

- In 2012, the United Nations Economic and Social Commission for Asia and the Pacific (UNESCAP) released *Low Carbon Green Growth Roadmap for Asia and the Pacific*. The roadmap provides "five tracks on which to drive the economic system change necessary to pursue low carbon green growth as a new economic development path" (UNESCAP 2012: 220).

All these conventional analyses seek, by and large, to improve the government's ability to overcome market failures while limiting the risks of resource misallocation and political capture to a minimum. However, as the market logic does not account for longer-term potential social and economic benefits, orthodox proposals have not been able to engineer solutions. In contrast, this contribution elaborates on thorough 'green industrial policies' and places these realistic strategic policies within the scope of the developmental state transformative involvement. A green industrial strategy would aim at identifying economic sectors, processes, and additional demand—through new consumption patterns, for instance—to be viewed as 'green'. Assuming that green growth requires capabilities that allow for handling complex, non-routine situations in both the government and the private sectors, and technological progress and industrial accelerators ought to be directed towards greener technologies to avoid squandering investment spending towards 'brown' technologies for short-term returns, developmental state action can promote structural economic change by altering polluting 'brown' sectors through intentional strategic industrial policy.

Within such a context, the main goal of this contribution is to link the developmental state view to green economic growth. This chapter comprises two major sections. Section 2 briefly reviews the developmental state literature while pointing out the absence of 'green considerations'. Section 3 frames a green developmental state by distinguishing two sets of factors: sociocultural and politico-institutional elements; and economic development and industrial targeting aspects. Finally, brief conclusions end the chapter.

2 Developmental State Theory in Retrospect: The Absence of 'Green'

The developmental role of the state has been mentioned intensively by a great number of authors. The idea of the developmental state had occurred long before it was defined and became a widely used economic development model in practice. In reality, developmental states, or more precisely, states with developmental functions, had even appeared long before they were named by scholars. For instance, the Netherlands in the sixteenth and seventeenth centuries; Britain from 1560 to 1851; and Germany from 1850 to 1914 are sometimes viewed as developmental states in certain aspects (Bagchi 2000). Marx, the most influential figure over ex-socialist countries, Vietnam and China, is also among the earliest to have a rudimentary notion of the developmental state. He is believed to have a basic notion of a capitalistic developmental state when mentioning the French state under Louis Bonaparte, which had thoroughly been strengthened against the civil society and had a "completely independent position" (Marx 1852: 238; Leftwich 2000: 155). Although this type of state emerged from the society's class structure and was seemingly able to insulate from any specific class interests, this state was not 'free-floating' as its basic goal was to strengthen the interests of the capitalist class (Leftwich 2000: 155).

In scholarly circles, the idea might be dated back to Friedrich List, an early theorist of economic protectionism, who is the first to counter contemporary prevailing liberalism by stressing national interests. However, the most persuasive account of late industrializing countries possibly belongs to Gerschenkron (1962) with 'the latecomer effect', which explained how late developers could 'catch up' with the developed world. Quite different to Marx's notion of industrial revolution and its prerequisites, Gerschenkron (1962: 8–10) conceives industrial revolution simply as cases of sudden considerable increases in the pace of industrial growth, and recognizes the 'formidable institutional obstacles' to the cause of industrialization. Getting closer to the concept of the developmental state are writers like Riggs, Jackson, and Huntington, who focused on the concentration of power in a 'bureaucratic polity'. Also, using a variant of the concept are theorists of the Dependency School, such as,

Cardoso and Faletto (1979), who utilized the notion of 'developmentalist state' applied to Mexico and Chile when these countries attempted to industrialize their economies during and after the war (Leftwich 2000: 156).

On the whole, by the end of the 1970s, although a large number of authors had analysed and investigated the developmental roles and functions of the states in different countries, no theorists had systematically conceptualized the developmental state. Only when Johnson's (1982) seminal work on Japan was published, was the concept of the 'developmental state' formally introduced and then seriously analysed by scholars and practitioners. Soon after that, the experiences of Taiwan and South Korea were also incorporated into the developmental state theory.

Pulling together the works of major contributors, a detailed developmental state analysis needs to encompass ten distinctive elements: (1) developmentally oriented leadership; (2) competent state bureaucracy; (3) pilot agency; (4) embeddedness; (5) sufficient state autonomy; (6) policy selectivity; (7) capacity to mastering the market; (8) ability to organize civil society; (9) competence to organize private interests; and (10) good performance and legitimacy.[1] The first five facets relate to the nature and quality of the organizational structure and personnel of the developmental state apparatus; the next two features illustrate its effective intervention mechanisms; and the rest refers to the outcomes of the first seven features of the developmental state model. These contributions clearly reveal the importance of long-standing heterodox threads, namely, the role of history, institutional structures, political economy, local culture and social psychology, and international relations, and place special emphasis on such important notions—including evolutionary-institutional ones—as emergence, path-dependence, power, idiosyncrasies, and national development acumen.

The element of 'green sectors and activities', however, has been missing from these heterodox writings. Green sectors can advance economic growth while ensuring that the quality and quantity of natural resources can continue to foster green investment and innovation, which would

[1] Johnson (1982), Amsden (1989), Wade (1990), Evans (1995), Leftwich (1995, 2000), Chang (1994, 2003), Karagiannis (2002).

underpin sustained development and give rise to new economic opportunities. On the policy front, what has been seen by successful developmental states, such as those in Japan and in China, is that the great emphasis placed on industrial growth targets may well have contributed to the neglect of adverse developments like environmental contamination, structural imbalances, and sociocultural evolution shortfalls. We talk about green growth which is nationally owned and transformative, enhances the capital domain assets, supports livelihoods and new opportunities, is sustainable and inclusive, and empowers people. These missing links from the developmental state literature will be filled in the subsequent section.

3 Towards a Green Developmental State Framework

Despite the enormous impact of processes associated with globalization and financialization, the developmental state remains a viable and influential feature of the development process. More to the point, the 'post-industrial era', intently coupled with hyperglobalization and financialization, called for a resurgence of industrial growth and revived a strong interest in 'go-back-to manufacturing' strategies and policies to achieve endogenous growth and better standards of living (Kaldor 1966, 1968). At the same time, imposed by global and national asymmetries in resource mobilization, rapid technological developments, market power and political influence, fresh interventionist approaches in selected industrial policy arenas have become once again attractive. Nowadays, the relevance and importance of industrial policy have been acknowledged by heterodox and mainstream economists and by different political leaders worldwide. While green growth initiatives are usually focused at the national level, they often fail to fully recognize that countries' institutions and policies can be at quite different starting points in the transition to a green economy.

Moreover, as an economy depends on its natural assets[2,3]—i.e., energy, biological, and mineral resources—which not only provide essential inputs, but also absorb residuals of economic activities—e.g., waste and emissions to soil, air, and water—economists, scientists, and decision-makers have placed special emphasis on the mechanisms and calibrated policy actions to foster the growth of specific 'green economic engines', that is, increase and strategic positioning of green targeted sectors, activities, and networks. Unlike resource-based economies, which depend critically on locations where natural endowments are abundant, a modern knowledge-based green economy depends on 'green capacity' and 'smart industrial specialization'. Government is the economic structure that is best positioned to make long-run investments. Subsequently, a green developmental state would need to provide alternative national visions and green national programmes to 'pick' or nurture green industrial sectors, perhaps even extending to specified firms and activities. In framing a green developmental state paradigm, two sets of notions are important: politico-institutional and sociocultural elements; and green development management and 'smart industrial targeting'. These sets of notions are analysed in the following two sub-sections.

[2] We extend the broad term of a country's 'natural assets' to also incorporate its 'blue economy', which comprises economic activities related to seas, oceans, and coasts, and covers interlinked emerging and traditional sectors (European Union, 2018). In addition to the long-established ocean activities such as tourism, maritime transport, and fisheries, the blue economy includes ocean ecosystem services—e.g., biodiversity, carbon sequestration, coastal protection, and waste disposal; economic externalities that are not accounted for by the private market. Besides, it encompasses areas where the private sector can take on a significant role, such as, aquaculture, bioprospecting, marine biotechnology, renewable energy, and seabed extractive activities (IBRD and World Bank 2017).

[3] The 'blue economy' has its origins in the 'green economy' concept, and is based on a sustainable development approach wherein developmental state involvement, similarly, supports targeted 'blue sectors' capable of promoting blue growth in a planned way. Such 'green' and 'blue' sectors and activities may well support endogenous development efforts to build resilient societies, and can form the foundation of a green or blue economy that safeguards sustainability and social and economic equity.

3.1 Sociocultural and Politico-Institutional Elements

Languages, religions, values, and customs differ from one country to another, which have been influenced by their history and have a profound impact on their economic development and policy outcomes. Yet, controversial changes that seek to transform existing structures and promote green growth can face severe challenges that may cause upheavals in the socio-economic milieu and thus may be politically problematic (Karagiannis and Madjd-Sadjadi 2007). To cope with these crucial challenges that can run contrary to the imperatives of 'instrumental' progress in any country and tip the debate in the direction of those social and political forces who seek to endorse a green developmental state approach, appropriate and supporting political, economic, and other social institutions will have to be in place.

Clearly, green economic development would require effective institutions grounded in norms of sustainability, accountability, equity, and tolerance for risk in pursuit of mutual gains for both the public and private sectors. In fact, a green developmental and entrepreneurial state would require wide consultation, broad political consensus, unwavering focus, a strong sense of realism, and realignment of decision-making with green development goals to facilitate the pathway to a successful production-oriented transformation. Furthermore, a green developmental state would need to provide ways for people, especially those who are excluded, to find a voice and influence in setting the green growth agenda.

In theory, Veblen's view—known as the 'ceremonial-instrumental dichotomy'—can provide an explanation of existing technological advances and production trends worldwide. Society progresses through learning how to deal with the material means of life. Thus, according to Veblen's analytical contribution, the key in human evolution is the invention of new, more effective green technologies, in our case. Veblen saw societies not only dependent on skills and tools to support the 'life process' but also having a 'stratified structure of status group life'. This gave rise to the dichotomy of the 'ceremonial'—related to the past—and the 'instrumental'—oriented towards the technological progress of the future (Veblen 1904; Knoedler 1997). Green economic development would involve transferring resources from established methods of production to green, innovative, welfare-enhancing methods.

Moreover, in line with Schumpeter (1934), the emergence of green complementary capabilities would develop around vital innovations to create green growth pathways. Such an alternative development model would generate a variety of diffuse economic and social effects that extend to complementary sectors. Following Schumpeter's analysis, green economic development would entail changing a county's industrial structure, the educational and occupational characteristics of its population and, therefore, its entire social and institutional fabric.

Such a radical alternative proposal seeks to overcome politico-institutional hurdles, is a long-term assignment requiring multidimensional changes, and would need a set of conditions favourable to such a policy shift. As green innovation is a social activity that requires different skills to collaborate to create value, the development of appropriate quality institutions is a major enabling factor behind green growth. High-quality institutions would support green productive activities and encourage capital accumulation, skill upgrading, invention, and green technological pathways. By investing in institutions, a green developmental state can support further utilization of endogenous industrial capabilities, lower risk, and inspire confidence by certifying the range of potential outcomes. However, as organizations and incentives flow from existing institutional arrangements, engaging in green growth means building or augmenting existing institutions that would be critical to progress. Building institutional capacities would also allow for a better platform to accommodate uncertainty and enhance the ability to meet possible contingencies.

Considering that an economy is built upon its own legacy, social values, codes of behaviour, and ethics, its policymakers must work with the country's stakeholders to develop a new partnership model for organizing and managing the relationships between the national government, regional authorities, and green industrial targets. As a well-articulated vision for government is crucial to pursuing a long-run course, green industrial targets can help a country's policymakers to have clear national development goals. To achieve this, a national government must bring together various interest groups representing the public and private sectors, as well as the broader society, to build consensus on vital and pragmatic strategic goals and objectives to be supported and executed by well-educated, well-trained, efficient, trustworthy, experienced, and morally committed technocrats. Executive bureaucrats and planners could form an entrepreneurial-type strategic planning agency that has the long-term commitment and the powers to intervene decisively and take necessary policy action (Karagiannis 2002; Karagiannis and Madjd-Sadjadi 2007).

Any government could benefit greatly from a planning team that has the capacity to coordinate green development priorities and industrial expansion initiatives, thorough technically proficient strategies, and a coherent set of policies. These planners need to be powerful members who have the necessary understanding of the required policy areas and sufficient knowledge of green technological pathways to 'create winners' and ensure a transition to ecologically sustainable trajectories. This presupposes clearly defined objectives which are broken down into measurable performance indicators. For government to be effective in promoting green economic development, there is a need for performance and impact assessments that are able to provide decision support for strategic investments. In fact, this requires monitoring and evaluation routines to continuously check the progress and performance of green industries and sectors, and support green programmes against existing benchmarks.

Thus, institutions will need to be in place to manage structural change, providing coordination for the emergence of new economic activities, nurturing entrepreneurship, and investing in education and skills development to ensure that human capital adapts efficiently to changes in the green aspects of the productive structure. As social elites may benefit from an existing unequal brown economy, unbundling the roles of policy design, funding, implementation, and evaluation can be very helpful to insulate such performance-based systems against any sort of collusive behaviour and interference. In addition, policymakers would need to maintain close relationships with the private sector and other stakeholders to get a deep understanding of how green economic sectors would function, and where bottlenecks exist that could hold back required changes. To what extent government intervention is necessary and what instruments are best suited are more likely to be context-specific and to change over time (Karagiannis and Madjd-Sadjadi 2007).

Consequently, government leadership and coordination will have to be accountable, effective, competent, and corruption-free to make the policy space for change, and this would require honesty and vigilance. This is especially important when few green sectors are the driving engines for green growth as exploitation of various loopholes and exemptions tend to appear more frequently when the stakes are higher. Still, the strategic planning process should be participatory and inclusive at all levels as participation would allow different social segments and national, regional, and local political forces to 'buy-in' to the goals and objectives that are clearly defined.

Green developmental state institutions will have to rely more heavily on complex collaborations that include government entities and programmes as important facilitators and conveners. The role of industrial policy could thus enhance embeddedness through systematic planning within broader social welfare processes that improve the outcomes for society at large. As the *nature* of a green developmental state would derive from more comprehensive state–society relations, special emphasis ought to be on expanding the space for public deliberation on the direction of green technological development as well as on policies of social inclusion to make sure that the society 'buys in'. In the absence of such participation, marginalized social and political forces and groups may refuse to support or go along with any controversial change or 'instrumental transformation'.

Clearly, the idea of 'green growth' has been promoted by various international institutions. But barriers to inclusive green growth are largely sociocultural and institutional in the broadest sense. Uncritical borrowing of these general analyses without the intervening stage of national institutions serves merely to import ideas which neither recognize the possibilities of change permitted by local conditions nor respect the limits on these possibilities imposed by them. As the Institutes which voice concern over 'green growth' are Western-oriented, genuine national efforts call for a broader cultural involvement in green development thinking, and propose a vision of a more humane and better society based on principles of equity and sustainability.

Ultimately, the best sustainable road ahead can only be found by way of analysis of the specific economic, social, and cultural conditions of any society. These multifaceted changes cannot occur without the complementarity or strategic partnership between the state and market participants, and without strong links with civil society, to transform an economy through targeted, dynamic, and promising green industrial sectors. Indeed, social and cultural patterns of behaviour would need to encourage green initiatives and engagement, as well as complementarity between government and industry. Without these preconditions and effective conflict management, *if* and *when* necessary, such a green developmental state framework will founder on short-term expedients, vested interests, socio-political challenges, or the resistance of 'ceremonial' arrangements and institutions.

3.2 *Economic Development and Industrial Targeting Aspects*

Enough evidence exists to show that the economies which are able to generate more effective long-term-oriented cooperative arrangements regarding (technological and organizational) learning and investment spending (in human and physical assets) are likely to outperform countries that largely rely on classic free market mechanisms. Developmental state involvement can channel investment into socially desirable directions, which may well aid green technological pathways and sustainable development. Accordingly, industrial policy can be conceived as a collaborative process of discovery and experimentation in which government and private actors closely interact and continuously negotiate and adapt their contributions to the development of green industries and sectors. Within the capital domain, there is a systematic inclusion of natural capital, human capital, social capital, and economic and financial capital, which all can influence sustainability and human well-being. Hence, policy interventions not only do they directly affect the capital domain but can also open up new economic opportunities.

As a snapshot of green industrial policies worldwide, Steer (2013) points out that the concept of 'green economic growth' can end up helping a national economy in more ways than just pollution reduction. Such policies can lead to competitive advantages that can create well-paying jobs and enhanced technological capabilities. Steer (2013) also mentions that over 50 developing countries are now implementing measures, such as feed-in tariffs and standards for renewable energy, "that at first sight seem not to be in their country's narrow interest" (Steer 2013: 18). Fankhauser et al. (2012) examine the resultant 'green race', which is seen as being not unlike that of the space race, in order to empirically determine what causes relative success across countries.[4]

However, modern economic development requires us to think about the dynamic and long-term trajectory of a country rather than concentrating on an efficient resource allocation in the short run. Consequently, using a green developmental state framework, instead of a *market failure* perspective coupled with institutional arrangements based on efficiency considerations, puts forward realistic alternative policy recommendations. Such a developmental state approach clearly requires targeting specific

[4] For an interesting orthodox contribution, see Rodrik (2014), 'Green Industrial Policy'.

green sectors, rather than taking a more broad-based approach to industrial policy, with the aim of greening endogenous production lines and reducing pollution levels.

Considering that part of production, consumption, income, investment, innovation, and labour may be identified as green, aggregate demand policies can contribute much towards improving any country's economic performance, and can facilitate green progress. As prudent macroeconomic management must seek to promote national purpose goals within budget constraints and to 'crowd in' private productive investments, fiscal policy along 'functional finance' lines—based on effective demand principles—is very important and relevant here, at least in the short and medium term (Lerner 1943). The 'functional finance' view would claim that whether budget deficits and national debt are too large ought to be assessed against the 'functional finance' benchmark, and would place special emphasis on the role of fiscal policy in supporting selected green dynamic sectors and industries and securing higher levels of economic activity. On the other hand, the objectives of monetary policy must maintain an interest rate policy that allows promising green industries to acquire necessary capital, encourage financial stability, and promote longer time horizons.

Crucial national purpose and internally specified goals will enable a nation's decision and policymakers to figure out local solutions to local needs and challenges. Such a green production approach would centre around a flexible and dynamic mixture of inward focus and outward orientation, and would require a technically competent government that has the ability and the teeth to engage in thorough, technically proficient planning. In addition, the mutual benefits among new agriculture, green manufacturing, and green services would boost forward and backward linkages, and would bring about diversification while developing a broad and more stable supplier base. Although new breakthrough advances emanate from R&D and skill formation, there are also green industries and modern forms of innovation that are within reach and rely on different areas of knowledge. More to the point, the development of the capacities of indigenous green sectors and firms can significantly expand a country's technological landscape and skills and may reduce the exposure to the vagaries of the world market forces and crises, thus, becoming more economically competitive in these areas in the near future (Karagiannis and Madjd-Sadjadi 2007).

The new targeted green areas, singled out by a country's developmental state, may need to be organized around certain indigenous firms of high potential and achievability that show the ability to dynamically address current challenges and would require substantial financial support. Green investment needs to be increased in both quantitative and qualitative terms, and linkages need to be strengthened so that development efforts and policy do not result in failure, exasperation, and waste of valuable resources on projects that could lock any economy into a low growth-inadequate skill mode for years to come. By focusing on the most promising economic development priorities that can take care of the human, material, financial, and politico-institutional requisites and direct local resources effectively, policymakers can create the necessary conditions for the successful implementation of a green growth pathway. The key is not to allow short-term issues and ad hoc solutions to hamper vital long-term parameters (Karagiannis and Madjd-Sadjadi 2007).

Rather than measuring economic success by the number of jobs created or higher growth rates, this alternative approach focuses instead on manufacturing value-added, overall productivity, and efficiency and distribution of gains among related industries, especially when natural resources have been facing certain constraints. Targeting and support of selected green sectors also require detailed information on the quantity (*how much*) and quality (*what type*) of modern factors, which are needed by these sectors so that the quantitative and qualitative parameters of planned industrial investment are thoroughly taken care of (Karagiannis 2002; Mazzucato 2013).

As capacity is essential to green investments and innovation, cautious development planning ought to match industrial policy goals with sustainability outcomes. However, to address a country's 'green capacity deficit', special emphasis needs to be placed on adequate green infrastructure and green industrial accelerators. As a result of this green innovativeness, manufacturing can create large knowledge spillovers, which could also enhance productivity in non-manufacturing activities. As 'traditional' financial schemes, such as loans or microfinance options, often recommend temporary assistance, may inspire squandering, favouritism, and rent-seeking, and can offer marginal solutions without getting at the root of the problems, there could be two useful fund-related options in the industrial policy arsenal that also highlight the long-term nature of green investments:

- Considering that jobs in green industries require specific skills, government-led development policy can address this 'green capacity deficit' with strategic infrastructural investments and improvements in human capital together with a special emphasis on finance and guidance of higher levels of investment and capital accumulation. A government can invest in those activities that the private sector does not find lucrative enough to warrant its own investment in the short term, or for which the capital requirements are large and the number of participants complexes that institutionalized policy intervention is required. Planned production-oriented spending on specific training programmes, skills, knowledge, R&D, research, and innovation can align finance with green industrial targets; increase the quantum of skills and expertise linking productivity improvements with incomes; help 'socialize risks and rewards'; and support green prioritized sectors and activities (Karagiannis 2002; Mazzucato 2013).
- Tax policy selectivity by sector, location, and ownership can be a useful tool in the industrial policy arsenal. Green taxes can broaden a nation's tax base, alter competitive advantages, and push markets into new dynamic areas because they can play a significant role towards focusing on industrial targets. However, these policy options need to emphasize performance, should be fund and time-limited, and would need effective systems of checks and balances. Discriminatory policy options through revenue enhancement or by channelling investment into targeted green sectors can be part and parcel of good industrial policy when cost discovery is at issue. Thus, discriminatory incentives and disincentives can not only transfer financial resources from disfavoured firms and activities (e.g., 'brown technologies') to favoured industries and sectors (i.e., 'green technologies') but can also release capital for green growth. Selective incentives and disincentives are reasonably impermeable to international challenge because they offer no clear advantage to the country using them. This has the extra benefit of simultaneously adding sectors that are being emphasized (Karagiannis and Madjd-Sadjadi 2007).

Furthermore, industrial policy tools, such as green industry protection measures, can reduce uncertainty by guaranteeing demand. As indigenous technological capabilities are the basis for a gradual correspondence between local production and demand structures, green investment priorities and the choice of techniques ought to be determined by the strategies of transformation and diversification, and by the product choices to which these strategies give rise. Sector targeting can help greatly in this regard and reflects a desire to support green economic growth. By accepting differentiation of key green sectors and industries, thorough industrial targeting can improve policy intervention effectiveness, strength, and evaluation while increasing rigour and transparency in government strategic decisions.

Such an industrial expansion will be characterized by green core industries around which linked industries develop, mainly through direct and indirect effects and positive spillovers. Even though reasonable production costs are expected, modern production techniques can make it possible to manufacture in small series on a viable basis. So long as firms are dynamic and swift footed and the developmental state is competent, trustworthy, and visionary, green industries can quickly achieve the necessary production levels to be globally competitive, and will result in improvements in both resource utilization and the balance of payments.

Yet, green industrial strategies could inevitably be most prone to open conflicts. This can be so because strategic industrial policy has to choose between technological pathways, sectors, or even individual firms in the same industry. Consequently, it tends to be more explicitly selective than other policies (Chang and Andreoni 2020). A green developmental state will give priority to those green sectors and industries, which appear viable and strategically important in a long-term perspective but which are vulnerable in the short or medium term without significant intervention. Such industries have to be provided with the resources and commitment to allow them to grow and mature, so that in the longer term the degree of intervention can be progressively reduced and the focus switched to other industries which have emerged as new priorities. As the green production base grows, and as an economy becomes more competitive technologically and on a human capital basis, additional opportunities for economic growth will come up.

Moreover, the formation of backward and forward linkages to the domestic economy could be a major problem. Evidently, externally enacted economic sectors not only cannot drive continuous structural changes and real production diversification but also exacerbate any country's exposure to the vagaries of the global environment. Instead, an economy should make a greater and better use of its productive resources and capacity as the expansion of indigenous production lines could result in a better and more egalitarian and sustainable industrial growth. Contrary to neoliberal thought, the state has an indispensable role in helping the development of green sectors of high potential and achievability that would drive planned structural changes through active industrial policy—thus, leading to the kind of overall industrial upgrading and diversification that sustains long-term growth.

Green industrial engines can open up possibilities and set up incentives for a wide range of new economic activities, and are expected to be supply chain partners for a country's other sectors, thereby resulting in more balanced economic structures. It is this process of 'smart specialization' and green innovation that could drive productivity growth and can determine where firms, regions, or countries have relative advantages. Furthermore, the government can emphasize selected 'green growth poles' as an important element of its strategic industrial policy. The expansion of the growth poles can bring about the expansion of output, employment, related investments and technical, technological, and psychological benefits based on anticipation of future demand.

In line with regional development and diversification efforts, and considering that industrial activity can benefit from location, a government can direct national resources in such green areas that have high growth potential. Location becomes important not only for recognizing opportunity but also for providing an environment that is responsive to the promising and dynamic industries. Such industrial targets may then venture into unchartered domains and can test the limits of their capabilities to realize potential gains. A proper green industrial policy needs to define the roadmap to success with timed benchmarks and incentivize the phasing out of unsustainable or highly polluting technologies. In other words, a committed national government needs to outline a thorough plan for action and undertake various initiatives to boost investments in green growth poles and industrial clusters.

Evidently, modern economies have overemphasized higher levels of foreign direct investment (FDIs) without serious consideration of any other plausible direction to activity. What's more, there are two very important industrial policy-governance related concerns: (1) careful studies have found little systematic evidence of technological and other externalities from FDIs (Hanson 2001); and (2) foreign investors often exploit various 'exceptions' and local resources while weakening a country's tax base (Cowling and Tomlinson 2011).

Although FDIs and joint ventures are policy options, in contrast to neoliberal proposals, an intelligent government pursuing a strategic industrial policy will not have a 'uniform' policy stance towards foreign investments across industries. Each industry serves different functions in the greater scheme of industrial development, and it would be totally unwise to have either uniformly liberal or uniformly restrictive policies towards FDIs across different industries. Instead, a more 'appropriate' approach towards foreign initiatives ought to be designed in the context of the long-term strategy for green development. What exactly the strategic way means depends on various crucial factors, such as a government's relative negotiating position, the technological nature of the industry, the role of the particular industry concerned in the bigger scheme of the country's environmental strategies and industrial diversification, and the broader social and economic needs of the nation (Chang 2003).

4 Conclusions

This chapter links the developmental state argument to green economic growth with a view that intervention, learning, adaptation, and fast and flexible experimentation in policy may never end. For governments, articulating an alternative vision and meeting a set of national development objectives is a challenging task, which depends on functioning social and economic institutions. A production-oriented approach that clearly prioritizes green development has been outlined, given the active role that national governments continue to play. More to the point, current conditions and trends can increase the potential advantages of pursuing 'governed-market' policies.

Looking forward, one would expect that sustainable development and green growth paths could signal the beginning of a new phase for the developmental state model. Assuming that green growth requires capabilities that allow for handling multifaceted situations in both the government and the private sectors, and technological progress and industrial accelerators ought to be directed towards greener technologies, a green developmental state apparatus can promote structural economic change through intentional strategic industrial policy. Due to its special focus on the economy–environment nexus, a green developmental state conceptually covers broader objectives and green policy responses, including green investments, green employment creation, green infrastructure, natural capital efficiency and sustainability, and vital aspects that are needed for human welfare and quality of life. Consequently, by finding the right focus, activities, and partners for inclusive and equitable green growth, such a green developmental state can pave the way towards a better society.

References

Amsden, A. (1989). *Asia's next giant*. New York: Oxford University Press.

Bacher, A. (2015). 'Task force on green economy: Ten conditions for a transition toward a green economy', Green Growth Knowledge Platform, September 4. https://www.greengrowthknowledge.org/resource/task-force-green-economy-ten-conditions-transition-toward-green-economy.

Bagchi, A. K. (2000). 'The past and the future of the developmental state', *Journal of World-Systems Research*, **6** (2), 398–442.

Cardoso, F. H. and Faletto, E. (1979). *Dependency and development in Latin America*, Oakland, CA: University of California Press.

Chang, H.-J. (1994). *The political economy of industrial policy*. London & New York: Macmillan & St. Martin's Press.

Chang, H.-J. (2003). *Globalization, economic development and the role of the state*, London and New York: Zed Books.

Chang, H.-J. and Andreoni A. (2020). 'Industrial policy in the 21st century', *Development and Change*, **51** (2), 324–351.

Cowling, K. and Tomlinson, P. R. (2011). 'Post the Washington consensus: Economic governance and industrial strategies for the twenty-first century', *Cambridge Journal of Economics*, **35** (5), 831–852.

European Union (2018). *The 2018 Annual Economic Report on EU blue economy*, Brussels: EU.

Evans, P. B. (1995). *Embedded autonomy: States and industrial transformation*, Princeton: Princeton University Press.

Fankhauser, S., Bowen, A., Calel, R., Dechezleprêtre, A., Grover, D., Rydge, J., and Sato, M. (2012). 'Who will win the green race? In search of environmental competitiveness and innovation', Centre for Climate Change Economics and Policy Working Paper No. 111, and Grantham Research Institute on Climate Change and the Environment Working Paper No. 94, September.

Gerschenkron, A. (1962). *Economic backwardness in historical perspective: A book of essays*, Cambridge, MA: Harvard University Press.

Global Green Growth Institute (2012). *Green growth knowledge platform*, Seoul, S. Korea: GGGI. https://www.greengrowthknowledge.org/organization/global-green-growth-institute-gggi.

Green Growth Knowledge Platform (2013). "Moving towards a common approach on green growth indicators: A Green Growth Knowledge Platform Scoping Paper", April. https://www.oecd.org/greengrowth/GGKP%20Moving%20towards%20a%20Common%20Approach%20on%20Green%20Growth%20Indicators[1].pdf.

Hanson, G. H. (2001). 'Should countries promote foreign direct investment?', G-24 Discussion Paper No. 9, New York and Geneva: UNCTAD.

International Bank for Reconstruction and Development/The World Bank (2017). *The potential of the blue economy: Increasing long-term benefits of the sustainable use of marine resources for small island developing states and coastal least developed countries*, Washington, DC: IBRD and World Bank.

International Chamber of Commerce (2012). *ICC Green Economy Roadmap—A guide for business, policy makers and society*, Paris: ICC.

Johnson, C. (1982). *MITI and the Japanese miracle: The growth of industrial policy: 1925-1975*, Stanford: Stanford University Press.

Kaldor, N. (1966). *Causes of the slow rate of growth of the United Kingdom*, Cambridge: Cambridge University Press.

Kaldor, N. (1968). 'Productivity and growth in manufacturing industry: A reply', *Economica*, **35**, 385–391.

Karagiannis, N. (2002). *Developmental policy and the state: The European Union, East Asia, and the Caribbean*, Lanham, MD: Lexington Books.

Karagiannis, N. and Madjd-Sadjadi, Z. (2007). *Modern state intervention in the era of globalization*, Cheltenham: Edward Elgar.

Knoedler, J. T. (1997). 'Veblen and technical efficiency', *Journal of Economic Issues*, **31** (4), 1011–1026.

Leftwich, A. (1995). 'Bringing politics back: Towards a model of the development state', *Journal of Development Studies*, **31** (3), 400–27.

Leftwich, A. (2000). *States of development: On the primacy of politics in development*, Cambridge: Policy.

Lerner, A. P. (1943). 'Functional finance and the federal debt', *Social Research*, **10**, 38–51.

Marx, K. (1852). *The Eighteenth Brumaire of Louis Napoleon*, New York City: Die Revolution.

Mazzucato, M. (2013). *The entrepreneurial state: Debunking public vs. private sector myths*, London: Anthem Press.

Organization for Economic Cooperation and Development (2011). *Towards green growth: A summary for policy makers*, Paris: OECD, May. https://www.oecd.org/greengrowth/48012345.pdf.

Rodrik, D. (2014). 'Green industrial policy', *Oxford Review of Economic Policy*, **30** (3), 469–491.

Schumpeter, J. A. (1934). *The theory of economic development: An inquiry into profits, capital, credit, interest, and the business cycle* (Harvard Economics Studies Vol. 46), Cambridge, MA: Harvard University Press.

Steer, A. (2013). 'Resource depletion, climate change and economic growth', paper prepared for the Global Citizen Foundation Project on 'Towards a Better Global Economy', February.

United Nations (n.d.). 'Green economy—Sustainable development knowledge platform'. Retrieved June 9, 2020, from https://sustainabledevelopment.un.org/index.php?menu=1446.

United Nations Economic and Social Commission for Asia and the Pacific (2012). *Low carbon green growth roadmap for Asia and the Pacific*, Bangkok, Thailand: UNESCAP. https://www.unescap.org/resources/low-carbon-green-growth-roadmap-asia-and-pacific.

United Nations Environment Programme (2008). *Green economy*, Washington, DC: UNEP. https://www.unenvironment.org/explore-topics/resource-efficiency/what-we-do/policy-and-strategy/green-economy.

Veblen, T. (1904). *The theory of business enterprise*, New York: Charles Scribner's Sons.

Wade, R. (1990). *Governing the market: Economic theory and the role of government in East Asian industrialisation*, Princeton, NJ: Princeton University Press.

World Bank (2012). *Inclusive green growth: The pathway to sustainable development*, Washington, DC: World Bank. https://openknowledge.worldbank.org/handle/10986/6058.

An Environmental Industrial Strategy for Sustainability

Malcolm Sawyer

1 Introduction

There have been many proposals for green new deals often focused on the funding of the investments associated with such deals and on reduction of carbon footprints.[1] The basic argument of this paper is that policies adopted to address the climate emergency and environmental degradation more generally should be embedded within an industrial strategy for environmental sustainability. The term "industrial strategy" may in some sense be an unfortunate one as the word industry may suggest the promotion of manufacturing and utilities which are often carbon-intensive and producer of harmful externalities of pollution. In this paper I'll use the term environmental industrial strategy.[2] Environmental sustainability is to

[1] For example, Fahnbulleh et al. (2019), Green New Deal Group (2008), Sanders (2019), Pollin (2021), European Commission (2019).

[2] I have elsewhere used the term social economic strategy (Sawyer, 2022), which is more encompassing.

M. Sawyer (✉)
University of Leeds, Leeds, UK
e-mail: M.C.Sawyer@lubs.leeds.ac.uk

N. Karagiannis and J. E. King (eds.), *Visions and Strategies for a Sustainable Economy*, Global Institute for Sustainable Prosperity,
https://doi.org/10.1007/978-3-031-06493-7_3

be widely interpreted to include the diverse ways in which carbon use can be reduced, and also attention to biodiversity and the general exploitation of the natural world.

In the next Sect. 2, I develop the case for an environmental industrial strategy. This follows the case of a strategic approach by government to industry in general, and it provides a framework for a comprehensive approach. In Sect. 3, the ways in which the investment required under an environmental strategy could be secured and specifically to ensure that there is a transition to a sustainable economy. Section 4 examines the ways in which macroeconomic policies can be used to support the environmental industrial strategy with regard to fiscal policies and to monetary policies. Section 5 offers some concluding comments.

2 The Case for Environmental Industrial Strategy

I take for granted that there has to be a wide range of policies, etc., to address the climate emergency and environmental degradation. The focus in this section is the argument for an industrial strategy approach to the environment.

The term "industrial strategy" has been used in a range of ways. A recent example is that of the UK government's short-lived industrial strategy (introduced in 2017, abandoned in 2020). The then Prime Minister, Theresa May, in her introduction to HM Government (2017), spoke of "at its heart it [the industrial strategy] epitomises my belief in a strong strategic state that intervenes decisively wherever it can make a difference. It is rooted in the conviction that a successful free-market economy must be built on firm foundations: the skills of its workers, the quality of the infrastructure, and a fair and predictable business environment" (p. 4).

Some have used the term "industrial policy" in a similar vein in a way which fits with what I would have termed industrial strategy, e.g. Chang and Andreoni (2020). The term "industrial policy" is itself used in many different ways. Industrial strategy goes beyond the traditional concerns of what many would term industrial policy—those of competition over monopoly, correction of externalities and public provision in specified areas. The particularly significant feature of industrial strategy is that it goes far beyond the market failure, control monopoly and promote competition approaches of neo-classical industrial policy.[3]

The use of the term strategy not surprisingly suggests taking a strategic approach (at the level of government) and also a form of industrial policy, and it draws in macroeconomic policies in its support including ensuring the funding of the investments required. The relationships between market forces and industrial strategy were laid out by Cowling (1990). "Just as there are systemic arguments ... for relying on market forces to play a centrally important role in modern economies, there are parallel arguments for imposing on these market forces coherent, community-based, national and supranational economic planning systems, within which they are allowed to operate." The case for an industrial strategy approach is reinforced when confronting the climate emergency. As Pistor (2021) argues "markets are an unreliable guide for navigating a problem as large and complex as climate change."

In Sawyer (2000), I wrote that what I termed the industrial strategy approach "does not lead to the advocacy of central planning: in part because of the informational and incentive requirements for successful planning are impossible to achieve. Instead, the government accepts a strategic role under which a broad view on future developments is evolved, and in which public support is forthcoming for productive activities (rather than exchange or financial ones). Much of the co-ordination of economic activity is undertaken through the market, though it is recognized that substantial parts of such co-ordination take place within firms and within households." The industrial strategy approach does not see a sharp dichotomy to be drawn between allocation through markets and allocation through planning.

The mainstream economics perspective had been to address environmental issues in terms of externalities imposed by production and consumption and the correction of those externalities through taxes and subsidies. However, in the context of the climate emergency, the issues to be addressed are systemic and there is limited usefulness of tackling externalities through taxes and subsidies.

Chang and Andreoni (2020) point to industrial policy tools which "reduce uncertainty by guaranteeing demand." These include infant industry protection which enables infant firms to survive and learn,

[3] For further elaboration on this see Sawyer (2000).

and restrict competition from foreign producers, limiting competition among domestic firms and government procurement policies. Further, "the government can provide a clear platform for technological evolution of an industry by taking a lead in the development of the basic technologies." Further, "the government can push firms to form research consortia to develop basic technologies, which they will share and use in developing more applied technologies." Of particular relevance for climate emergency: "during the early stage in the development of an emerging industry, where different technological standards compete with each other, the government can reduce uncertainty about the path of future technological evolution by imposing a technological standard ... the government can [also] subsidize or directly provide technology-related 'public goods' (such as data, metrology, prototyping and testing facilities)." These lines of argument are all clearly highly relevant to tackling climate change and environmental degradation.

An industrial strategy is viewed as government-led mapping out the developmental path for an economy, from which supporting industrial, labour and macroeconomic policies are derived. An environment industrial strategy would identify sectors, processes and consumption patterns to be viewed as environmentally friendly. As such it is a wide ranging set of policies—it is not just the promotion of renewal energy, reduction of air travel, but influencing consumption patterns, addressing loss of biodiversity. Industrial policies are then required to foster the development of industrial processes and products which contribute—the most obvious example in recent times would be renewable energy in its different guises, recognizing that the social benefits of such processes exceed the private benefits. Further, the pay-offs from investment in such processes are subject to fundamental uncertainty and long-term, and "lock-in" effects are significant.

The European Commission (2019) advances a European Green Deal (EGD) which should serve as a growth strategy to "transform the EU into a fair and prosperous society, with a modern, resource-efficient and competitive economy where there are no net emissions of greenhouse gases in 2050 and where economic growth is decoupled from resource use" (p. 2). The EGD organized around eight interlinked macro-areas for action: Increasing the EU's climate ambition for 2030 and 2050; supplying clean, affordable and secure energy; mobilizing industry for a clean and circular economy; building and renovating in an energy and resource efficient way; accelerating the shift to sustainable and smart mobility; designing a fair, health and environmentally friendly food system; preserving and restoring ecosystems and biodiversity; a zero pollution ambition for a toxic-free environment (taken from Fig. 1 of EU, 2019). There is also the stated intention to mainstream sustainability into all EU policies with explicit references to the EU budget, macroeconomic and fiscal policies, taxation, research and innovation, education and training, trade policies and EU external policies. As Pianta and Lucchese (2020) argue "the integration of a green industrial policy with Europe's overall policy framework is a key requirement for success. In fact, a joint approach to environmental and industrial policy has close connections with fiscal, monetary, competition, trade, technology, and labor and wages polices that have to be considered" (Pianta and Lucchese, 2020). They further argue that in facing "a green transition, market-based processes are expected to work poorly," and that "the transition to a sustainable economy requires deep changes in technologies and production systems" (p. 637). "The phasing out of old technologies and the introduction of new sustainable ones has major, widely differing effects on firms, industries, and workers across regions; the distribution of the benefits and costs should be considered, and appropriate adjustment policies should be developed" (p. 637). These considerations all point in the direction of the requirements for an environmental industrial strategy.

A crucial element of an environmental industrial strategy comes from the decisions over which projects, investments, sectors to support and develop and by implication which are not to be so supported. It involves the re-structuring of the economy in an environmentally sustainable direction. There are hard choices to be made and noting the "fundamental uncertainty" which surrounds these decisions. This requires, inter alia, mapping out the shifts in the structures of economic activity which would be required, the identification of the types of investment which are consistent with addressing the climate emergency and environmental degradation. A crucial feature of an environmental industrial strategy would be setting out what is regarded as "green," and what is not, and the degrees of "greenness." Terms such as "green investment" (and "brown investment") are often used to signify in a dualistic manner those investments which are deemed to be environmentally friendly on some criteria. A government formulated and transparent assessment of what would constitute "green investment" is required to inform the government's own funding policies, the central bank's asset purchase policies and also as guidance for those seeking to support "green investments" through their own savings.

These types of policies face the formidable difficulty of defining and then monitoring those investments which may be considered "green," low carbon, preserving nature and biodiversity, etc. EU (2020) provides an example of seeking to draw up a taxonomy of "green investments." It also has to recognize that the assessment of whether an investment should be deemed "green" has to take into account not just the parts of a production process to which the investment relates but also other parts of the overall chain of production. It is then highly relevant that the pursuit of selective credit policies be undertaken within the context of a "climate emergency industrial strategy" which formulates the criteria on which "green investment", etc., is to be judged, and which sets up arrangements to monitor the application of the criteria. There are though weaknesses from labelling bonds as "green bonds." "The current system of green bond labels does not necessarily guarantee a material reduction in carbon emissions. Indeed, these labels would signal emission reductions only if the relevant projects were to transform the activities of the bond issuer radically enough for its carbon emissions to fall." However, the authors show that "green bond labels are not associated with falling or even comparatively low carbon emissions at the firm level" (Ehlers et al., 2020, p. 31).

There are clear difficulties in ensuring that finance is flowing into environmentally friendly projects. It is required to consider the full range of impacts of a specific investment, for example where the investment is part of a supply-chain. As such the project would draw on other production processes which themselves may be environmentally damaging. In some instances, the uses to which the products which will be produced will be put and the ways in which the product may be re-cycled or may be dumped have to be taken into consideration.

3 Funding an Environment Industrial Strategy

In considering the funding of investments in an environmental industrial strategy, there has to be acknowledgement of the volume of funds for investment which would be required elsewhere in the economy. In industrialized economies there has been a tendency for slowing of economic growth (of GDP) which many would see as exacerbated by the global financial crises of 2007/09 and the COVID-19 pandemic of 2020/21.

Li and Mendieta-Muñoz (2020) ask (in terms of the title) "are long-run output growth rates falling?", and answer that their results "show a permanent reduction in long-run output and technical progress growth rates that is not associated with the detrimental effects of GR [great recession]. ...These findings suggest that the slowdown in productivity—and not other factors associated with labour force growth—has been the main driver of the decline in long-run GDP growth" (p. 19).

Many would see through the lens of secular stagnation that slower growth stretching into the future. Others, including those within the de-growth school, would see the necessity for slower, near-zero rates of growth of GDP to address global warming. In general, slower growth would involve lower levels of investment in so far as the capital-output ratio does not rise. From that perspective, an environmental industrial strategy may not need higher rates of investment and its funding unless "green" investments are much more capital intensive than others or if the overall rate of growth of GDP comes out faster than in recent years.

The lower growth and indeed decline of carbon emissions regarded as vital to avoid global warming above 1.5 °C, which requires a combination of slower growth of GDP and the de-coupling of carbon emissions from GDP. The transition to a low-carbon sustainable economy would involve a shift of resources including capital equipment from carbon-intensive industries into low-carbon industries. Lewney (2021) argues that "because the net zero technologies are more capital intensive, the overall scale of investment in the energy system would be higher, especially in the period to 2050 when the entire new system needs to be put in place." However, he points out that the estimates of the level of investment which would be required vary quite widely and depend on the ways by which the emissions reductions are achieved. For example, investment in energy efficiency may have lower investment requirements than investment in alternative energy sources. "For a 1.5 °C pathway, IPCC (2018) draws on existing studies to present an average estimate that annual global investment in the energy system amounting to $2.4 trillion (at 2010 prices) would be needed between 2016 and 2035, equivalent to 2.5% of world GDP. This is a gross figure, meaning that the energy system would in any case require substantial investment in a current-policies baseline case (about 1.8% of global GDP (IPCC, op. cit., p. 373), but the additional investment is still substantial." Each of these lines of argument would suggest that the need for investment in productive capacity would be reduced. In so far as the capital requirements for production (goods and services) are broadly proportional to output (constant capital-output ratio) then slower growth involves slower growth of capital stock, and thereby a lower investment to GDP ratio. From that perspective there is a lower requirement for the funds for investment purposes. It can also be noted that there is a tendency for governments to run budget deficits, and the counterpart of that is that private domestic savings plus borrowing from abroad are greater than investment. A transition to a low-carbon economy would require substantial investments in appropriate "green" investments; but it would require reductions in investment in industries which are highly polluting and damaging to the environment. The general perspective here is that a "green industrial strategy" may well not need an overall increase in investment and saving, but rather the re-direction of investment and saving. Thus, the focus has to be on the allocation of finance and funding rather than on raising the overall levels of saving and investment.

First, consider the role of the banking system. A more significant element is that the financing of investment, initial and final in the terminology of the circuitist approach, needs to be channelled towards socially desirable and environmentally friendly investment, and not towards financial asset accumulation and speculation. The thrust of the operations of the banking and financial system should be on the financing and funding of real investment, and not financial investments.

Pollin (1998), writing in the context of financial structures and egalitarian policies, noted that "finance is the conduit for all economic activity in market economies. Because nothing happens unless it is financed, exerting control over the financial system is an efficient way to influence the widest possible range of activity with a set of relatively small and simple policy tools" (p.163). In respect of financial institutions, he argues that "bank-based systems are better equipped [than market-based systems] to promote longer time horizons and a stable financial environment. Their structures also create more favourable conditions for activist government interventions, including both traditional macro policies and public credit allocation policies" (p. 164). This line of argument can be broadened out in a number of respects. It has to be recognized that the differences between bank-based and market-based systems are one of degree in that most financial systems have both banks and financial markets. Indeed, banks are the major routes through which money is created (and destroyed) in the process of loans being provided, and it is difficult to envisage a financial system without banks. The term "banks" though covers a wide range of financial institutions including what may be termed commercial banks, some of whose liabilities (bank deposits) are treated as money with close relationships with the central bank. There are also savings banks, investment banks, etc., which are deposit accepting and in general regulated. Within the banking sector, the patterns of ownership vary from the private profit-seeking, mutual and co-operatives through to the State owned. Mutual financial institutions have often been more focused on households—for example housing finance (as in the UK's building societies) and credit unions.

Policies of selective credit allocation may provide one route through which "green investment" can be encouraged. Many countries operated policies of selective credit allocation in the first decades after World War II, though such policies have fallen out of favour in recent decades. Selective credit allocation requires banks to allocate finance and funds towards specified priority sectors—at the limit only to those sectors, and at other times requiring a high proportion of lending to the priority sectors. "Virtually all central banks have engaged in "industrial policy" or "selective targeting." The difference lies in which industries they have promoted. Significantly, the whole tenor of economic development can be fundamentally affected by which of these industries the central bank and associated institutions promote" (Epstein, 2007). Epstein further indicates the wide range of ways in which credit controls have been operated.

The roles of State development banks (with names such as Green Investment Bank) are of particular importance. As Griffith-Jones and Cozzi (2016) argue, "the existence of development banks is justified by the existence of sectors and investment projects that require funding for the future development of the economy, but have high uncertainty as to their future success." These projects may find difficulty in securing private funding because of the uncertainties involved. The long-time horizons, the fundamental uncertainty of investment projects, and the path dependence involved are often particularly important with the climate emergency.

Development banks "have a double mandate. They mainly aim to maximize sustainable and inclusive development impact (including economic, environmental and social impacts), while maintaining some financial profits or avoiding financial losses," contrasted with commercial banks "whose main (and usually only) aim ... is to maximize risk adjusted expected returns, often short-term ones. ... While commercial banks need to manage the range of economic, environmental and social risks, they generally only do so to the extent that these risks impact financial returns" (Griffith-Jones et al, 2020). They argue that public development banks should support long-term productive investments, including those that emphasize low-carbon projects, as well as those benefiting poorer regions and populations, and base their selection on criteria that put development impact first, with financial returns an important but secondary objective. The strategic role of the PDBs is stressed by "PDB should aim to shape the future, and move from being mere 'project-takers' to 'project-makers,' playing a proactive, first-mover role to help overcome uncertainties and risks, and define missions, programs and projects."

Griffith-Jones and Carreras (2021) look to ways in which European Investment Bank (EIB) "has important achievements in improving the quality and directionality of the investment it funds towards a far greater focus on the low-carbon economy, but … it must do even far more to improve this directionality, given the scale and the urgency of the challenge of mitigating and adapting to climate change. It must move even further from being just a project taker (responding to demand coming from the private sector) to also project maker … and thus help in creation of new markets and sectors, such as for example that of hydrogen, so promising especially if generated competitive from green energy."

In a green deal approach (and in an environmental industrial strategy), there will be projects which would be environmentally friendly and socially beneficial but which do not readily generate financial returns. There are activities which need to be de-commodified and taken outside of the market (e.g. provision of universal basic services). Finance Watch (2020) argue for the role of what they term public finance as "being especially useful for projects that provide public value without being financially bankable themselves," in particular with regard to protection of nature and biodiversity. It points out that "the nature of private investment is to focus on financial risks and returns and revenue streams, but **many nature-related projects have no revenue source**. Indeed, nature tends to benefit when there is less economic activity. Conversely, it can be easier to finance businesses that keep their costs low by harming nature" (bold in original). It argues that "**Nature projects are often too small for institutional investors to invest in directly**. They are complex to understand, illiquid and take a long time to mature. …. the [finance] industry's reliance on CAPM and other structural factors mean that fund managers have little appetite for alternative or long-term investments."

4 Macroeconomic Policies for an Environmental Industrial Strategy

Any industrial strategy needs to be supported by fiscal and monetary policies. In general, this has meant a set of macroeconomic policies supportive of a sustained level of demand and elements of public expenditure, including investment consistent with the industrial strategy (Arestis and Sawyer, 1999). It is necessary to go beyond that view in the case of an environmental industrial strategy, and consider some of the components of fiscal and monetary policies, and how those policies need to be expanded.

4.1 Fiscal Policy

The underlying principle for the overall fiscal policy is the use of the budget deficit position to secure a high and sustainable level of employment and output (Sawyer, 2019a, 2019b). Here I put to one side issues on working time and the degree to which rising productivity would be taken in terms of shorter working time. Further, I continue to take full employment as an objective of policy, recognizing the difficulties of defining full employment; and without exploring policies such as universal basic income which could change what is deemed to be full employment. In the context of an environmental industrial strategy, the public expenditure programmes have to be designed accordingly. The focus of public investment would itself be on environmentally friendly infrastructure, particularly of the form which does not yield a financial return. On the tax side, the structure of tax would shift towards taxes on environmental "bads." Taxes such as wealth tax, higher corporation tax, property/land taxes would help with an equality agenda, and can have positive effects for environmental policies in so far as the rich as much greater polluters than the poor.

4.2 Monetary and Central Bank Policies

Central banks should pay full regard to climate change in their decision-making, but that there would be rather limited consequences of that. The mandate of a central bank should be formulated away from inflation targeting, and towards concern over financial stability and more importantly here that monetary policy is conducted in a manner fully consistent with addressing climate change and environmental degradation. The UK Chancellor of the Exchequer has announced (March 2021) there is to be a reformed mandate for the Bank of England to "reflect the importance of environmental sustainability and the transition to net zero." The role of the central bank is though likely to be a rather limited one, and specifically limited to purchase of financial assets deemed to be "green" in the context of QE and the purchase of corporate bonds.

The policies of QE conducted by central banks over the past decade have raised a number of policy suggestions. First, since QE appears to provide money to the banking system through purchase of financial assets then could not money created by the central bank be supplied to others? Second, the central bank decides which financial assets it will purchase and which it will not, and the quantity of financial assets to be held. It has traditionally been the case that a central bank engaged in open market operations or similar with purchase/sale of government bonds.

The operations of the central bank should be pursued in ways consistent with the broader political and social objectives of the government—in this context, that is it should be a decision of government as to what is regarded as "green investment" and the degree of "greenness." In effect, a central bank being willing to purchase/hold some financial assets but not others shifts the composition of demand for the financial assets, and to that extent the price of "green" financial assets is somewhat higher than otherwise, and the price of "brown" financial assets somewhat lower. The effects could be similar to those of the disinvest movement (which then raises the question of who decides which type of assets are to be favoured and which not). Dafermos et al. (2020) provides an example from the Bank of England of how the present arrangements can have adverse effects in respect of climate change. "[I]n its present guise [in 2020], the CBPS [Corporate Bond Purchase Scheme] is mis-aligned with the government's climate goals and implicitly creates better financing conditions for carbon-intensive economic activities. The CBPS biases the allocation of capital towards carbon-intensive sectors, while at the same time failing to reflect climate-related financial risks."

Climate change impacts on the financial system through two main channels of risks. One can be described in terms of the physical risks from damage to property, infrastructure and land. The other set of risks come from "changes in climate policy, technology, and consumer and market sentiment during the adjustment to a lower-carbon economy" (Grippa et al., 2019, p. 26). "Financial stability concerns arise when asset prices adjust rapidly to reflect unexpected realizations of transition or physical risks, there is some evidence that markets are partly pricing in climate change risks, but asset prices may not fully reflect the extent of potential damage and policy action required to limit global warming to 2 °C or less" (Grippa, et al., 2019, p. 27–8). More generally, "Financial institutions could incur losses on exposures to such firms with business models not built around the economics of low-carbon emissions. These firms could see their earnings decline, businesses disrupted and funding costs increase due to policy action, technological change and shifts consumer and investor behaviour. Risks can materialize especially if the shift to a low-carbon economy is abrupt (as a consequence of prior inaction), poorly designed or uncoordinated globally. Going forward, a key next step in developing stress tests for transition risks will be to capture 'second-round' effects—in which a decline in asset prices leads to fire sales, which further depress asset prices, generating a vicious cycle and an amplifying mechanism for an initial shock."

Reduced global demand for fossil fuels is likely to lead to "stranded assets" with loss of profits on the underlying assets and falls in market valuation of the assets. Mercure et al. (2018) analyse the macroeconomic impact of stranded fossil fuel assets. "Our analysis suggests that part of the SAFA [stranded fossil fuel assets] would occur as a result of an already ongoing trajectory, irrespective of whether or not new climate policies are adopted; the loss would be amplified if new climate policies to reach the 2° target of the Paris Agreement are adopted and/or if low-cost producers … maintain their level of production … despite declining demand; the magnitude of the loss from SFFA may amount to a discounted global wealth loss of US$1–4 trillion."

There can be issues of financial instability arising from the challenge of climate change. There is recognition that there is what has been termed "transition risks" arising from the re-valuation of carbon-intensive assets arising from shifts to a low-carbon economy.[4] The change in asset values (assumed to be downwards though the valuation of carbon-unintensive assets could well rise) has implications for the range of financial institutions and households who own the corresponding assets. This may well be another example of financial markets mispricing financial assets—why have the risks involved not been incorporated into the financial asset prices?

[4] For example, Breeden (2019), Dafermos et al. (2018).

While it may be relevant for the central banks and others to warn about the likely shifts in asset prices, it is far from clear what actions would follow for monetary policy.

5 Concluding Comments

There are urgent needs to tackle the climate emergency and environmental degradation as is now widely recognized. In this chapter I have argued for an approach which may be labelled a comprehensive environmental industrial strategy. An industrial strategy is viewed as government-led mapping out the developmental path for an economy, from which supporting industrial, labour and macroeconomic policies are derived. An environmental industrial strategy would identify sectors, processes and consumption patterns to be viewed as environmentally friendly. Industrial policies are then required to foster the development of those environmentally friendly industrial processes, recognizing that the social benefits of such processes exceed the private benefits. An industrial strategy approach is required to formulate policies which shifts resources and employment into "green" activities, and also mitigate the adverse effects of shifts of employment away from environmentally damaging activities. An industrial strategy approach recognizes not only the strategic roles of government but also the limitations of "market forces" in addressing the climate emergency. The pay-offs from investment in environmentally friendly processes are subject to fundamental uncertainty and long-term and 'lock-in' effects are significant. The environmental industrial strategy has to be supported by the funding of environmentally friendly investment, and policies such as a public investment bank have significant roles to play, particularly in light of fundamental uncertainty and lock-in effects. The environmental industrial strategy has to be supported by macroeconomic policies. The underlying principle for the overall fiscal policy is the use of the budget deficit position to secure a high and sustainable level of employment and output. Central banks should pay full regard to climate change in their decision-making, but that there would be rather limited consequences of that. The mandate of a central bank should be formulated away from inflation targeting, and towards concern over financial stability and conducted in a manner fully consistent with addressing climate change and environmental degradation.

References

Arestis, P. and Sawyer, M. (1999), 'The macroeconomics of industrial strategy' in K. Cowling and R. Sugden (eds.), *Industrial Policy in Europe*, London: Routledge, 1999, pp. 352–370

Breeden, S. (2019), 'Avoiding the storm: climate change and the financial system', Speech given at Official Monetary & Financial Institutions Forum, London 15 April 2019.

Campiglio, E. (2016), 'Beyond carbon pricing: the role of banking and monetary policy in financing the transition to a low-carbon economy', *Ecological Economics*, 121, pp. 220–230

Chang, H-J and Andreoni, A. (2020), 'Industrial policy in the 21st century', *Development and Change* 51(2), pp. 324-351

Cowling, K. (1990), "A new industrial strategy", *International Journal of Industrial Organization* vol. 8, pp. 165-183

Dafermos, Y., Gabor, D., Nikolaidi, M. and van Lerven, F. (2020), 'Decarbonising the Bank of England pandemic QE: perfectly sensible?', *Policy Briefing*, New Economics Foundation, June 2020

Dafermos, Y., Nikolaidi, M., and G. Galanis, (2018), 'Climate change, financial stability and monetary policy', *Ecological Economics*, 152, pp. 219-234

Ehlers, T., Mojon, B., and Packer, F. (2020), "Green bonds and carbon emissions: exploring the case for a rating system at the firm level", *BIS Quarterly Review*, September 2020, pp. 31–47

Epstein, G. (2007), "Central banks as agents of economic development" in H.-J. Chang (ed.) *Institutional Change and Economic Development*, United Nations University and Anthem Press, pp. 95–114

EU Technical Expert Group on Sustainable Finance (2020), *Financing a Sustainable European Economy: Taxonomy*, Brussels: European Union

European Commission (2019), "The European Green Deal", Brussels 11.12.2019, COM(2019) 640 final

Fahnbulleh, M., Powell, D., Pendleton, A., Stirling, A., Van Lerven, F., and Balata, F. (2019), A green new deal in the UK. London: New Economics Foundation. Retrieved from https://neweconomics.org/2019/04/a-greennew-deal

Finance Watch (2020), *Nature's Return: Embedding environmental goals at the heart of economic and financial decision-making* (Authors Ludovic Suttor-Sorel, Nicolas Hercelin)

Green New Deal Group (Simms, A., Pettifor, A., Lucas, C., Secrett, C., Hines, H., Leggett, J., Elliott, L., Murphy, R., and Juniper. T.) (2008), A Green New Deal: Joined-up policies to solve the triple crunch of the credit crisis, climate change and high oil prices. London: New Economics Foundation, on behalf of the Green New Deal Group. Retrieved from https://neweconomics.org/2008/07/green-new-deal

Griffith-Jones, S. and Cozzi, G. (2016), 'The roles of development banks; how they can promote investment, in Europe and globally' Chapter 5 in Noman, A. and Stiglitz, J. E. (eds) *Efficiency, finance, and varieties of industrial policy: guiding resources, learning, and technology for sustained growth*, New York: Columbia University Press, pp. 131–155

Griffith-Jones, S., and Carreras, M. (2021), *The Role of the EIB in the Green transformation*, Brussels: The Foundation for European Progressive Studies (FEPS)

Griffith-Jones, S., Spiegel, S., Jiajun, X.U., Carreras, M. and Naqvi, N., (2020), "Matching risks with instruments in development banks", AFS Paris Research Paper

Grippa, P., Schmittmannm J., and Suntheim, F. (2019), 'Climate change and financial risk', *Finance and Development*, December 2019

HM Government (2017), *Industrial Strategy: Building a Britain fit for the future*

IPCC, (2018), *Global Warming of 1.5°C. An IPCC Special Report on the impacts of global warming of 1.5°C above pre-industrial levels and related global greenhouse gas emission pathways, in the context of strengthening the global response to the threat of climate change*. Intergovernmental Panel on Climate Change.

Jackson, T., and Victor, P.A. (2019), "Unraveling the claims for (and against) green growth", *Science*, vol. 366 issue 6468, pp.950-1

Lewney, R. (2021), 'Environmental Policies to save the planet' in P. Arestis and M. Sawyer (eds.), Economic policies for a post neo-liberal world, Palgrave Macmillan, pp. 179–224

Li, M. and Mendieta-Muñoz, I. (2020), "Are long-run output growth rates falling?", *Metroeconomica*, 71(1), pp. 204-224

Pianta, M. and Lucchese, M. (2020), "Rethinking the European Green Deal: an industrial policy for a just transition in Europe", *Review of Radical Political Economics*, 52(4), pp. 633-641

Pistor, K. (2021), "Green markets won't save us", *Social Europe* 24th March 2021

Pollin, R. (2021), "The industrial policy requirements for a global climate stabilization policy", *International Review of Applied Economics*, 35(3-4), pp.389-406

Pollin, R., (1998), 'Financial structures and egalitarian economic policy' in P. Arestis and M. Sawyer (eds.), *The Political Economy of Economic Policies*, Houndmills: Macmillan, pp. 162-201

Sanders B. (2019), Sanders and Ocasio-Cortez announce the green new deal for public housing act. Press release. Nov 14th. 2019. Retrieved from https://www.sanders.senate.gov/newsroom/press-releases/sanders-and-Ocasio-Cortez-announce-the-green-new-deal-for-public-housing-act.

Sawyer, M. (2000), 'The theory of industrial policy' in *New Challenges to Industrial Policy*, Kluewer Academic Publishers, edited by Wolfram Elsner and John Groenewegen, pp. 23–58

Sawyer, M. (2019a), "Approaching budget deficits, debts and money in a socially responsible manner" in P. Arestis and M. Sawyer (eds.), *Frontiers of Heterodox Macroeconomics*, ISBN 978-3-030-23928-2, Palgrave Macmillan. pp. 45–87

Sawyer, M. (2019b), 'The Fiscal Responsibilities of Government' in Nikolaos Karagiannis and John E. King (eds), *The Elgar Companion to Modern Government Intervention*, Cheltenham: Edward Elgar, Chapter 3, pp. 85–96

Sawyer, M. (2022), "Beyond social democracy and neo-liberalism: towards a social economy", PSL Quarterly Review, 74,299, 287-306

Sustainability and Economic Policies

Proper Future Economic Policies

Philip Arestis and Nikolaos Karagiannis

1 Introduction

The focus of this contribution is on proper future economic policies so that economies that utilize them would be able to produce high levels of economic activity and employment, along with equal distribution of income and wealth. The economic policies suggested are fiscal, monetary and financial stability policies but most importantly, coordination of these policies to achieve the relevant goals is paramount. We refer to the US, the Economic and Monetary Union (EMU) and the UK as necessary.

In order to do so, we begin with the experience of economic policies before and after the Great Financial Crisis (GFC) of 2007–2009, and during the more recent Covid-19 syndrome. This is followed by our

P. Arestis (✉)
University of Cambridge, Cambridge, UK
e-mail: pa267@cam.ac.uk

N. Karagiannis
Department of Accounting, Economics and Finance, Winston-Salem State University, Winston-Salem, NC, USA
e-mail: karagiannisni@wssu.edu

N. Karagiannis and J. E. King (eds.), *Visions and Strategies for a Sustainable Economy*, Global Institute for Sustainable Prosperity,
https://doi.org/10.1007/978-3-031-06493-7_4

third section, which concentrates on the necessary changes to introduce the policies and coordination of them to achieve the goals of high levels of economic activity and employment. The final section summarizes and concludes.

2 Pre- and After–GFC and Covid-19 Economic Policies

Pre-crisis monetary policy was not successful. The Inflation Targeting (IT) type of policy, which had been introduced by many countries prior to the GFC was not the cause of price stability over that period (Angeriz and Arestis, 2008). It is also the case that the models used by the Central Banks prior to the GFC, failed to predict the crisis. Following the GFC, Keynesian economic policies enabled the avoidance of turning into a second Great Depression. In addition, many Central Banks responded aggressively by providing credit to the financial sector, increasing bank reserves, undertaking Quantitative Easing (QE) policies and keeping the focus on interest rates. Both monetary and fiscal policies were utilized in the USA and in the UK. It is the case that fiscal policy has primarily limited the impact of the Covid-19 syndrome; but not in the EU, where only monetary policy was implemented.[1] In the Euro Area the fiscal rules did not allow proper use of fiscal policy (relying only on automatic stabilizers, which are the endogenous part of any fiscal system, and account for around 30% for the advanced economies, with their results having "little credibility", as reported in Acocella et al. (2016, p. 181).[2] However, more recently they were relaxed in view of the Covid-19, and for only two years—2020 and 2021. The IMF has urged the Euro Area countries' government spending to increase by an extra 3% of their GDP (as reported in the *Financial Times*, 15 April 2021). This is to mitigate the impact of Covid-19. This proposal is due to the IMF's view that the Euro Area countries are expected to recover slowly in relation to other

[1] It should be noted that the institutional dimension of the Euro Area needs careful re-design so that its federalist structure is completed, and relevant policies are in place, and properly employed (Arestis, 2019a; Acocella et al., 2016).

[2] As Acocella et al. (2016) also suggest "Automatic stabilizers can fail either because they are too limited or because they are not powerful enough in oiling the market mechanism during large recessions, especially when restricted by fiscal rules designed to maintain sustainability" (p. 288).

countries. Ehnts and Paetz (2021), suggest that national governments should make extensive use of their regained sovereignty. The Stability and Growth Pact (SGP) should be reformed since increasing the debt to GDP ratio is not a problem, especially so when unemployment is high. Not surprisingly then, the Euro Area has moved into a recession in the first quarter of 2021. Its GDP dropped by 0.7% in the fourth quarter of 2020, and 0.6% in the first quarter of 2021, thereby two consecutive quarters of negative growth emerged as the technical definition of recession suggests. This is mainly due to coronavirus lockdown measures. The Euro Area is lagging behind the USA and the UK.

After 2010 austerity policies were introduced, which did not help at all. Austerity stagnation followed because policy interventions were very weak and due to governments introducing austerity in 2010. Austerity policies are supposed to remove excess fiscal deficits and reduce public debt. This is based on an undervaluation of the size of the fiscal multiplier. With the fiscal multiplier being greater than one, deficit reduction leads to a larger loss of GDP and thereby rising public debt and the debt ratio.[3] No wonder, King (2016), the ex-Governor of the Bank of England, suggested that austerity policies imposed "enormous costs" where they were implemented at that time. Post-GFC, the zero bound of interest rates has not helped. In addition, low interest rates have produced a great deal of unsustainable household debt.[4] Indeed, the Bank for International Settlements (BIS; as reported in the Guardian, 03 December 2020) suggested that it was similar to the pre-2007 era. McCausland and Theodossiou (2016) examine whether fiscal contraction could produce recovery in the public debt. They study theoretically and empirically the impact on public debt of fiscal policy in the case of eleven OECD countries over the period 1881 to 2011. Their conclusion is that fiscal contraction deteriorates public debt as a percentage of GDP. Increase in government expenditure reduces the public debt,

[3] Acocella et al. (2016) provide a short review of fiscal multipliers to conclude that they exceed one in the downturn. Thereby, "austerity policies would be counterproductive in recessionary periods, making public deficits and debt worse rather than better" (p. 166).

[4] In the UK, however, household wealth has risen during the pandemic period. The Office for National Statistics reported that this increase emerged from a rise in house prices of 8.5% in 2020, and of government support (as reported in the *Financial Times*, 30 April 2021).

and unemployment. A recent and relevant example is the UK's government borrowing. It is expected that the economic recovery after the Covid-19 syndrome should rapidly improve government borrowing, and thereby reduce the public debt, according to *Financial Times* relevant analysis (as reported in the *Financial Times*, 21 May 2021). Austerity, therefore, deteriorates rather than enhances economic activity. Skidelsky (2017, pp. 226–269) suggests that Keynes would have analysed the GFC and would have made the following proposals. Relevant Central Banks should have limited credit to the housing sector prior to the GFC, and similarly with the asset market in the case of the pre-great depression period. In addition, Keynes would have proposed the Central Banks rescue operations of 2009. However, Keynes would not have proposed the premature suspension of the QE in 2010. Keynes would not have supported the austerity policies introduced in 2010; instead, Keynes would have proposed increases in aggregate demand. It is also the case that excessive fiscal austerity holds back productivity, as the case of the recent UK experience has demonstrated (Wren-Lewis, 2020). It is also the case that in the UK the Covid-19 syndrome, along with Brexit, has contributed to the weakening of the economic activity more than in other countries.[5] The IMF suggests (as reported in the *Financial Times*, 08 April 2021) that advanced economies debt levels are expected to stabilize after Covid-19. Countries' heavy borrowing in 2020 contributed to better economic outcomes; low interest rates also helped countries to borrow, and the expectation is that the public debt would stabilize by the middle of the current decade.

However, the UK growth rate is expected to be 7.9% for the three quarters of 2021, after the first quarter, due to the speed and efficacy of the strong vaccine programme (the Bank of England expects the growth to be 7.25% in 2021).[6] In addition, lockdown has pushed savings higher, to 16.1% of disposable income (as reported by the Office of National Statistics), a near record, in the fourth quarter of 2020 (as

[5] In terms of the Brexit, an independent commission has been set up (April 2021) to closely examine, and improve, the UK's deals with the EU. This commission includes members of Parliament and business leaders (*Financial Times*, 12 April 2021).

[6] Tenreyro (2021) suggests that " In Europe, employment support schemes such as the UK government's furlough scheme mean that governments temporarily take over wage bills to avoid mass unemployment" (p. 6).

reported in the *Financial Times*, 01 April 2021), which enhances household consumption at the end of stricter lockdowns. Thereby recovery would be enhanced, especially in the post-Covid period; and as Haldane (2021) suggested, these higher savings "would provide a very significant degree of additional demand to an already rapidly-recovering economy" (p. 18).[7] In fact, the UK economy is recovering better than anticipated, and the unemployment rate is falling due to the reasons mentioned above and to fiscal support. It should be noted, though, that Brexit and in particular adjusting to it, and to changed trading arrangements with the EU, would produce some challenges. But at least, the economy is stronger to help manage these challenges. The UK unemployment is faced with the problem of "uneven" recovery. The most worrying aspect is that recent graduates and in the relevant job markets there is continuing lack of jobs for graduates (*Financial Times*, 05 May 2021). Would such a UK economic situation produce inflation, is an interesting and relevant question. The answer is both pessimistic and optimistic. Pessimistic in that the higher forecast growth rate would produce inflation. Optimistic in that the GDP expansion will be kept in check in view of economic uncertainty of the future. Also, shortage of supply in view of the lockdowns in 2020 and 2021, should disappear after the lockdowns, restricting current inflation spike, thereby inflation becoming transitory. However, the annual UK inflation rate increased in May 2021 to 1.5% from 0.7% in March 2021, due to price increases of raw materials. This increase in the UK inflation rate is expected to exceed the 2% inflation target by the end of 2021. The rate of inflation, though, that excludes prices of raw materials rose from 1.1% in March 2021 to 1.3 in April 2021; and pay awards have been at 1% since November 2020 (reported in the *Financial Times*, 20 May 2021).

Other policies emerged: QE whereby large-scale asset purchases by Central Banks are undertaken. The US Fed started undertaking QE in late November 2008 and the Bank of England in early 2009.[8] The

[7] Savers in the UK are assumed to have accumulated £180bn in their bank accounts (10% of UK's annual GDP). If all this were to be spent, it would increase consumption about 6% in both 2021 and 2022 (*Financial Times*, 12 April 2021).

[8] The Bank of England and the Treasury, and on 17 March 2020, established the Covid Corporate Financing Facility (CCFF), a lending facility. The aim of the CCFF is to provide lending to large non-financial firms. The Treasury decides which firms can take part in the CCFF.

ECB announced in January 2015 to buy public sector debt and private sector assets; this was a succession of more cautious interventions.[9] These "unconventional" monetary policies have now become the "new conventional", and as Driffill (2015) suggested, "QE usefully cuts the burden of serving the public debt and eases the government budget constraint" (p. 19). It should be noted that these "unconventional" and "new conventional" policies have been around after the GFC, and have continued now with the "Covid-19 syndrome". And as Bailey et al. (2020) suggest, the relevant policies after the GFC were "unprecedented", more recently with the Covid-19 syndrome they have become "more familiar parts of the Central Bank toolkits" (p. 1).

In terms of the transmission of the QE channels, Bailey et al. (2020) suggest three relevant groups: "First, those associated with policy 'signalling'. Second, those associated with the 'portfolio balance channel'. Third, those associated with market liquidity" (p. 6). The ultimate objective is to lower long-term interest rates, stimulate spending, increase GDP and influence inflation, the ultimate objective, so that it reaches the relevant target (Bailey et al. 2020, p. 8). In terms of the ultimate objective, unconventional policies have not been effective, and of course, these macroeconomic effects are the most important for policymakers (Bailey et al., op. cit., p. 9; see, also, Goodhart and Ashworth, 2012; and King, 2019). However, and in terms of the empirical and theoretical research since the GFC, suggest that QE has successfully lowered long-term interest rates, providing the required monetary stimulus. The evidence on the impact of QE in terms of market dysfunction and illiquidity has been the largest (Bailey et al., 2020). It is also the case that when QE, and other "unconventional policies" were introduced and implemented, had positive effects in terms of financial stability, increased the liquidity of the financial sector and reduced short- and long-term interest rates; they

[9] The ECB at its meeting on 21 January 2021 decided to leave its stimulus policies unchanged, and to check further whether bank lending rates, corporate credit conditions, and government bond yields "are favourable enough to boost demand and inflation" (reported in the *Financial Times*, 22 January 2021). However, further clarification and details of how 'favourable financing conditions' is defined are necessary. In any case, a further problem for the ECB is the recent sell-off in bond markets, which is spread to the Euro Area from the USA (where the expectation of higher inflation in view of the expected economic recovery has caused it). As a consequence of this sell-off, the ECB has increased its emergency bond-buying programme to avoid it. The Euro Area bond yields actually did fall.

were thus "somehow successful" (Goodhart and Ashworth, 2012).[10] A problem with the enormous QE is that it could push investors into the cryptocurrencies. This would make the policymakers very concerned in terms of the size of the market and their unchecked activity.

Forward Guidance (FG) has also been introduced. FG intends to influence private sector expectations so that they are consistent with the Central Bank's policy intentions in relation to interest rates, and thereby influence investment and growth. This is also all very relevant to the Zero Lower Bound (ZLB) interest rates. FG is expected to influence bond yields, whereby Central Banks state the time path for short-term interest rates, or any precise relevant conditions, such as until targets are reached, or for an indefinite period. Central Banks may suffer from reputational risk if they do not undertake what they initially intend to do. FG to be effective needs to be credible.

Another conventional monetary policy instrument is "funding for credit". This is to enable financial institutions to have access to credit from their Central Bank when the supply of credit is otherwise restricted, and as long as they meet specified lending objectives. The relevant Central Bank provides collateralized funding so long as banks meet specified lending objectives (introduced in the UK in 2012).

It should be noted that conventional monetary policy is expected to stimulate the economy, when the Central Bank buys short-term government bonds in order to lower short-term interest rates and operates through its overnight rate to the interbank market. When policy rates hit the ZLB, or close to it, the relevant transmission mechanism no longer works. Unconventional monetary policy is thereby introduced whereby the Central bank operates on long-term interest rates, in an attempt to lower them by large-scale asset purchases (QE) or indirectly by FG.

Post-GFC, bond yields have been persistently low in terms of both nominal and real terms. This can be explained by several factors. Falling productivity growth, declining tax rates on capital income, and the demand for high-quality liquid assets resulting from new regulations; and of course due to QE. More recently, the USA long-term bond yields reached their highest level since February 2020 (*Financial Times* 09 February 2021). This was due to the expansionary fiscal policy announced

[10] A further QE type, suggested by ecologists, is for Central Banks to provide free funds for green projects. Or finance at a zero interest rate, green bonds from firms engaged in ecological projects.

by the then USA President, which investors expected to increase the level of economic activity and inflation. In addition, and on 8 February 2021, the 30-year Treasury yield, and for the first time since February 2020, traded above 2.0% and 2.52% in March 2021, the highest level since 2019 (*Financial Times*, 19 March 2021); also, USA bond yields traded about 0.1% below, early April 2021, its 14-month high of 1.78%. All this was due to the current USA President's $1.9tn economic stimulus bill, which was passed by Congress in March 2021.[11] Similar increases in government bond yields have emerged in the 0.9trK and the EMU (not as high in the EU as in the USA and the UK). Under current circumstances, such increases in bond yields could induce central banks to implement targeting of the bond yields as part of the "yield curve control". Still though, long-term government bond yields remain low by historical standards.

Financial stability has convinced policymakers of its importance, especially so in view of the fact that the GFC demonstrated the limits of price stability in terms of ensuring financial stability. A relevant question is the extent to which the GFC crisis could have been avoided by relevant policies, such as financial regulation. In terms of financial stability, and a financial regulatory authority separate from the Central Bank, but in strict cooperation with it, is a solution with many advocates (see, for example, Stiglitz, 2010). Still, a great deal is needed in terms of its relationship to monetary policy, and its implementation.

In terms of the impact of unconventional monetary policies, Haldane (2016) suggests six channels as follows: policy-signalling channel, which is a form of FG; portfolio balance channel; liquidity premium channel; exchange rate channel; confidence/uncertainty channel; and bank lending

[11] More recently, however, middle of April 2021, Treasury bond prices have increased, in view of higher demand for Treasuries, since investors are uncertain about the impact of the $1.9tr impact on the economy; high inflation is expected. The 10-year Treasury bond rate of interest decreased to 1.55%. This, however, may change if the $1.9tr impact on the economy is successful. USA's consumer-price index (PCI) was 2.0% in March 2021. It increased from the 1.7% in February 2021, the biggest increase since 2009. It is expected to increase to 3.5% by May 2021 (*The Economist*, 17 April 2021). Whether a similar increase of the Fed's price index is consistent with the Fed's change of its monetary policy framework, as in August 2020, is an interesting question, since no clarification by the Fed on this issue, implies further uncertainty. The April 2021 surge in inflation to 4.2% (the highest annual rate since 2008) produced selling of the 10-year government bonds, with the relevant interest rate increasing to 1.68% (*Financial Times*, 13 May 2021).

channel. Unconventional monetary policies are not without problems. Financial instability may very well occur in the long run. Unconventional monetary policies could lead to excessive risk taken by insurance companies and pension funds, which need long-term assets to match their long-term liabilities. Inequality is also affected, which emerges in view of lower interest rates, which benefit borrowers over savers, and the portfolio rebalancing whereby investors hold more equities than bonds producing wealth effects. There is, of course, the question of the relationship between Central Bank unconventional policies and inequality. Colciago et al. (2018) provide relevant results from extensive theoretical and empirical aspects to conclude that, in general, and to the extent central bank policies stimulate economic activity, income inequality decreases. However, such policies by increasing asset prices increase income inequality. In terms of wealth inequality, higher financial asset prices increase inequality; higher housing prices produce lower wealth inequality. There is also the question of the impact of macroprudential financial stability policies on inequality, which can have similar effects. For example, in the case of collateral requirements, and as Stiglitz (2015) suggests, "lowering of collateral requirements ..... does not result in an increase in the overall efficiency of the economy, but leads to more inequality" (p. 27). Similar results are also relevant in relation to the rest of macroprudential policies. In terms of relevant empirical evidence on the impact of macroprudential policies on inequality it is scarce; therefore, reliable empirical results are not in evidence, and "it remains unclear through which mechanism macroprudential policy may generate distributional effects" (Colciago et al., 2018, p. 23).

When the rate of interest was reduced to zero nominal rate, Central Banks expanded high-powered money, the monetary base, through QE. The broad money stock hardly changed and bank lending to the private sector stagnated, and in some cases, it fell. As a result, the money multiplier simply collapsed. The main reason being that banks increased their deposits with the Central Bank instead. Banks cannot increase their lending if demand for it from the private sector is weak, as it is when expectations of future economic activity are weak. In addition, when banks reduce their lending in view of bad debt expectations, as, for example, in the Euro Area during the Covid-19 syndrome and relevant lockdowns, and during the GFC (*Financial Times*, 20 January 2021). What is needed under these circumstances is fiscal policy in coordination with monetary policy. It should be noted, though, that the EU countries have ensured support to those mostly affected by the Covid-19 pandemic. Also, the EU "Next Generation EU" recovery fund, financed by issued EU debt, commonly with the EU 27 member states, and the fund of $750bn to be provided in grants, aims at raising growth and improves long-term fiscal sustainability. Governments that need short-term funds to stimulate their economies, can borrow to spend currently. As reported in *The Economist* (15 May 2021), the $750bn recovery fund is to be implemented later in 2021 and expected to produce the EU's GDP to its pre-pandemic level by the end of 2021. This scheme is not large enough for the whole of the EU, but for the countries in need it is significant (Italy and Greece are two relevant examples).

In the Euro Area, countries overall have not had effective effect of the ECB unconventional monetary policies. The impact on output is weak, while the effect on inflation is insignificant. The harmonized consumer price rose to 1.3% in March 2021 from 0.9% in February 2021 (*The Economist*, 03 April 2021); lower than the ECB target inflation rate. Core inflation, which excludes energy and food prices, fell from 1.1% to 0.9% (*Financial Times*, 01 April 2021). There have been asymmetrical effects across countries. Countries that were affected by the crises have done badly, while countries that were not affected seriously by the crises have done better (Elbourne et al., 2018). The results of Elbourne et al. (op. cit.) are very similar to other studies on the Euro Area experience as reported therein. In any case, the ECB did not change its main interest rate until July 2008; other Central Banks had already changed their main interest rates well before 2008.

Commercial banks, and in view of very low interest rates due to unconventional monetary policies and sluggish economic activity, tend to hold excess reserves with their Central Banks. Some Central Banks have decided to charge commercial banks for their excess reserves; these are reserves above the minimum reserve requirements with Central Banks. Negative interest rates therefore emerge, which are expected to lead banks to provide more funding to the private sector, thereby enhancing investment. Negative interest rates may cause currency depreciation, thereby producing higher domestic demand. However, negative interest rates may cause commercial banks to buy assets instead, thereby putting upward pressure on the asset prices and borrowing costs. This could produce financial instability, which "creates costs in terms of bankruptcy of both financial and non-financial firms, the reduction of GDP and of asset prices" (Acocella et al., 2016, p. 255). Financial instability may also occur in view of bank profitability erosion. Non-bank financial institutions may also be faced with funding problems that could produce financial instability. Banks and non-bank investors may take excessive risks in their search for positive yields. This could produce asset bubbles, and financial instability. It is the case that Negative Interest Rate Policies (NIRPs) have been introduced by some Central Banks (as for example the ECB, the Swiss National Bank, the Danish National Bank, and the Swedish Riksbank). IMF (2015) suggests that "A number of major central banks in Europe have set key policy rates at negative levels in order to further encourage lending by making it costly for banks to hold excess reserves at their central banks" (p. 3). In addition, IMF (2017) considers the case of NIRPs to conclude that such policies as implemented in several Central Banks have had small positive effects on domestic monetary conditions. Prolonged periods of NIRPs, could produce a switch of deposit holders from their deposits to cash, which would have serious effects on banks' profits, and would imply bank difficulties in extending credit, with serious negative implications that would produce financial instability. According to the IMF (2017), "clear theoretical guidance is compounded by the fact that at there is little past experience to provide guidance" (p. 5) on the impact of NIRPs. In view of the limited impact of NIRPs policies, such as fiscal policy, structural and financial sector reforms, would be necessary. The Bank of England is concerned with the negative interest rates in view of the uncertainty over its impact. The current Governor of the Bank of England, as reported in the *Financial Times* (18 May 2020), said that the UK Central Bank was not contemplating negative interest rates since it would cause problems for the banking sector in terms of putting pressure on banks' profit margins. It could actually have more "harm" than "good". Also, and as reported in the issue of the *Financial Times* (18 May 2020), the USA Fed has a similar position on this issue. How monetary policy would influence the economy under negative interest rates is not clear enough. As reported in the *Financial Times* (05 February 2021), the governor of the Bank of England said that the Bank " did not wish to send any signal that it intended to set a negative bank rate at some point in the future". It should also be noted that negative interest rates have distributional effects in that creditors lose out and debtors benefit. It is also the case that countries that introduced negative interest rates have not produced the expected increases in spending and investment. An interesting comparison in terms of recent interest rates is between the USA and the Euro Area. Negative interest rates in the Euro Area have not been helpful. In the USA, the yield curve is steepening, in view of the Fed's effort to keep short-term policy rates low, with expectations of inflation and high growth rates pushing long-term interest rates higher. This is profitable for the financial sector in that it can get funds at low interest rates and lend at high interest rates, which helps the sector to provide loans for investment.

It is the case that following the GFC, Central Banks have taken a range of actions to avoid the collapse of the financial institutions and markets. Prudential regulation and supervision of banks, building societies, credit unions, insurers and major investment entities, all of them have been under these actions of the Central Banks in order to reduce systemic risks. Thereby, in addition to monetary and microprudential policies, which had been around prior to the GFC but have been strengthened since the GFC, and macroprudential policies have been introduced; these policies influence monetary and financial conditions and can therefore have an impact on monetary and financial stability (see, also, Bank of England, 2015). It is crucial that policymakers communicate on the risks of financial stability and account for how they influence monetary policy. Our suggestion is that coordination of monetary and macroprudential is of paramount importance in this respect. The post-GFC era has clearly shown that these policies along with the unconventional QE, the Central Bank buying government and other types of bonds and achieving the Zero Lower Bound (ZLB) of interest rates, and FG, did work. These recent policies, only at the beginning of their introduction and implementation, managed to produce a sustainable financial sector, but have not helped subsequently to enhance significantly the GDP growth rate and influence the inflation rate as required. Fiscal policy is very important in this respect, along with coordination of monetary and macroprudential policies.

Monetary financing whereby the Central Bank buys government bonds creating thereby base money, which can be used to finance fiscal policy, is another type of QE. This type of QE and fiscal policy can be used in two ways, namely, the government uses the base money created to reduce taxes, or to increase government spending. This, however, needs cooperation between the Treasury and the Central Bank. This type of policy and cooperation is possible, and had been employed before the introduction of the Central Bank's independence (Goodhart, 2012). However, it cannot be accepted since then in view of the "Ricardian Equivalence Theorem" of the New Consensus Macroeconomic (NCM) model part of the focus of this theoretical framework is on Central Bank independence. There are of course serious problems with the NCM theoretical framework (Arestis, 2019b; Arestis and González Martinez, 2015); and according to Muellbauer (2020), NCMs are "insufficiently dynamic, stochastic, general equilibrium, new, or Keynesian" (as reported in Vines and Wills, 2020, p. 481; see also, Blanchard, 2018; and Wren Lewis, 2020). Also, the NCM failed to account for what had been happening in terms of the economic events prior to the GFC, especially its failure to capture the asset prices boom, and, in addition, it failed to suggest policies that would have helped economies to recover from the very slow recovery subsequently (see, also, Vines and Wills, 2020, p. 428, footnote 2). Also, and as Stiglitz (2012) suggested, "What is especially remarkable is that central banks had models in which banking did not play an important role. In their own self-interest, they should have tried to make banking important. And banking is important, even though their own models did not capture them" (p. 40; see also Arestis, 2019b).

It is the case, nonetheless, that QE policies have a type of fiscal character and this implies that there is some relationship between governments and Central Banks. Coordination, therefore, of fiscal and monetary policies is possible and important. Coordination of fiscal and monetary policies should be helped by the creation of an independent fiscal policy authority (Arestis, 2020; Acocella et al., 2016). Such a fiscal policy council that sets sustainable fiscal policies is relevant in terms of the coordination of policies. Such a fiscal policy council "could review the fiscal outlook, estimate current structural imbalances, and give advice on correcting those imbalances if the fiscal position looks likely to become unsustainable" (Acocella et al., 2016, p. 290). A relevant example is the UK's Office for Budget Responsibility, which provides independent fiscal forecasts and their implication in terms of growth and employment. It is also the case that QE and other unconventional policies, do not affect growth and inflation as required, in which case fiscal policy is required, especially when coordinated with monetary policy.[12]

In the case of QE monetary authorities purchase significant quantities of long-term Treasury securities thereby raising their prices and

[12] In the EMU, the independent Central Bank and the lack of EMU fiscal policy does not allow proper coordination of fiscal and monetary policies. Proper fiscal union is required to allow such a proper coordination of fiscal and monetary policies in the EMU (see, also Arestis and Sawyer, 2011; Sawyer, 2013).

lowering the real rate of interest (their yields, the effective interest rate earned by holders of bonds). Investors, who sell Treasury securities, invest their money to other securities, thereby lowering long-term interest rates. The Central Bank affects in this way the spread of long-term interest rates over expected policy rates (what is called term premium); also the required return on risky assets relative to risk-free assets (what is called risk premium). All this is expected to influence interest rates and thereby consumption and investment. Clearly then, unconventional policies are of two types. QE and Credit Easing (CE). QE refers to injections of money to commercial banks from the Central Banks. CE is when Central Banks provide liquidity directly to the economy by buying private sector assets, such as corporate bonds and mortgage-backed securities. These policies are expected to increase the broad money supply, increasing asset prices, thereby producing wealth effects; thus, decreasing borrowing costs and stimulating expenditure. Influencing inequality is another possible effect, in view of their effect on income distribution. All these effects could produce long-run financial instability in view of lending could produce risk taking, especially to those companies that rely on long-term assets. FG is another relevant policy, whereby Central Banks explicitly announce future policy trends. This is particularly the case when the ZLB level of interest rates is approached in view of limited effectiveness of monetary policy when such low level of interest rates is achieved. Under such circumstances not only FG but also further unconventional policies are introduced as noted above.

Central Banks have also introduced stress tests. The Bank of England has introduced stress tests to determine the extent to which banks and building societies need to build up buffers of capital if their current ones are not large enough to avoid a crisis. The Financial Policy Committee (FPC) decides how severe the results of the tests are and thereby suggests the extent to which banks and building societies are able to absorb losses and continue to lend (it should be noted that the FPC runs macroprudential policy alongside the Monetary Policy Committee).

The economic situation after Covid-19 is expected to be accompanied by "a great uncertainty" (Bailey, 2021, p. 4). Unemployment is not expected to decrease substantially; economic policies are thereby very important. Bailey (op. cit.) suggests that such "economic policies would increase public and private debt, which should be managed by higher trend rate of growth, boosted by stronger investment" (p. 7). The Bank's Financial Policy Committee should make sure that sufficient supply of finance for productive investment should be in place (Bailey, 2021, p. 8). Increasing supply capacity is also important for it raises "the sustainable rate of growth of the economy and thereby ease the task of managing the necessarily higher level of public and corporate sector debt" (Bailey, 2021, p. 2).

The IMF Managing Director, Kristalina Georgieva, suggested in her statement at the G20 Ministers and Central Bank Governors (2020), "the outlook for global growth for 2020 is negative – a recession at least as bad as during the global financial crisis or worse". However, recovery in 2021 is expected; and added, "We strongly support the extraordinary actions many countries have already taken". In addition, "We welcome the moves of major central banks to ease monetary policy".[13] It should be noted that active fiscal, financial stability and monetary policy coordination has not been properly introduced; and it should. It is the case, therefore, that the GFC and the global Covid-19 crisis and economic policies undertaken and implemented by governments and Central Banks call for rethinking of economic policies for the future, especially coordination of relevant policies.

3 Necessary Economic Policy Changes

There are suggestions that the IT should change from 2 to 4% to make it easier for Central Banks to achieve negative real interest rates, to combat recession as necessary.[14] But IT's theoretical background has a

[13] As reported in the IMF Press Release No. 20/98 (23 March 2020).

[14] It should be noted that the Fed's inflation targeting is different from other central banks' (Central Banks follow the Taylor Rule, whereby they manipulate their interest rates to achieve their inflation targets, and also they respond to fluctuations to the output gap). The Fed has two objectives: 'maximum employment' and 'stable prices' (with 2% its inflation target). In terms of the 'maximum employment', which is interpreted as low unemployment, there is the question of the extent to which maximum employment is equivalent to 'low unemployment' (Tarullo, 2017, p. 3). Another difference is that the Fed has recently promised to tolerate higher inflation than its target (what is called, 'average inflation target', as stated in the *Financial Times*, 10 and 15 May 2021; with a number of the ECB's governing council members supporting the Fed proposal for the ECB). This implies that the Fed is prepared to see the real rate of interest reduced, thereby increasing investment and aggregate demand. Whether such a policy can be successful depends heavily on expectations of healthy future economic activity, which are around now in view of the President's $1.9tn economic stimulus. A relevant development is that in March 2021 the Fed upgraded to 6.5% the growth rate of the US economy, higher

lot of problems as suggested in Arestis and Angeriz (2008) and Arestis and González Martinez (2015). A relevant IMF Working Paper (Balima et al., 2017), using a meta regression analysis (MRA), and utilizing 8059 estimated coefficients from 113 empirical studies on the macroeconomic effects of IT, concludes that published empirical results suffer from publication selection bias, and no significant impact of IT on inflation volatility and real GDP growth has been identified. Targeting nominal income through interest rate manipulation is a better way forward; but it should be coordinated with financial stability and fiscal policies, as argued below. Fiscal policy should also focus on public debt management, especially unwinding of QE. Indeed, "Coordination calls for discussion among policy agencies before actions are taken, and willingness on the part of each policy to take account of the reactions of the other before committing itself to particular actions" (Sinclair and Allen, 2017, p. R9). Orphanides (2017) suggests that in the case of the Euro Area monetary and fiscal policies have not been effective in view of their tightness, which has contributed to further divergence among the Euro Area members. It should be noted that the Euro Area is only a Monetary Union, with no fiscal policy as part of the union since there is no central Euro Area budget. Fiscal policy is entirely with the individual members of the union. Fiscal Union is required, which would help to enhance the economic prospects of the Euro Area. Other contributors have also suggested coordination of fiscal and monetary policies. Lesper (1993) was right to argue that "Analysing one policy is like dancing a *tango solo*: it's a lot easier, but it is incomplete and ultimately unfulfilling" (p. 3; see, also, Acocella et al., 2016, Chapter 13). In addition, Dixit and Lambertini (2001, 2003) emphasized coordination between monetary and fiscal policies and used the term "Symbiosis" for policy coordination. Our suggestion is that proper coordination of fiscal and monetary policies, along with financial stability ones is required. This would also need a Fiscal Union.

than the December 2020 relevant rate of 4.5%. It also stated that it expected to achieve its 2% inflation target, which however it should go back to the target. However, the Fed is expected to keep interest rates close to zero until 2024. By contrast, the Governor of the Bank of England told the House of Lords Economics Affair Committee that if inflation moved above the 2% inflation target of the Bank of England, the Monetary Policy Committee would raise interest rates – but there is not much evidence of this at present (reported in the *Financial Times*, 19 May 2021).

Another problem with the ECB, and as reported by Orphanides (2017) is its reliance on private credit rating agencies in its determination of the eligibility of country government bonds for the ECB QE. This eligibility involves unequal effects in terms of its treatment of countries' government bonds since the ECB does not operate QE the same way for all Euro Area countries. This implies divergent effects among the Euro Area countries.

A relevant comment in the case of the Euro Area is on its SGP, which does not really help. It has recently been temporarily abandoned in view of the Covid-19 syndrome, until March 2022, but it should be abandoned forever, especially so if and when Fiscal Union emerges, when fiscal and monetary policies would need to be coordinated and with the financial stability.

Further changes, and for the period after the GFC, have been suggested by Blinder et al. (2017), based on views expressed by Central Bank heads and economists, from relevant surveys. These include the following: financial stability "considerations should also be part of the central bank mandate" (p. 1). In terms of monetary policy instruments, macroprudential instruments and FG should also be used. Central bank communication, which has been stepped up recently should be maintained and even increase it further. Central bank independence is also supported in this context. These views are strongly supported in that "in the future, central banks will operate with broader mandates, employ a wider range of policy tools and place even more emphasis on their communication" (p. 4). Yellen (2014) also suggested that financial stability, and macroprudential part of it, is an important aspect of financial policies; Indeed, "the global financial crisis and, more recently, the euro crisis have reminded us that no economy is immune from financial instability and the adverse effects on employment, economic activity, and price stability that financial crises cause" (p. 1). An additional suggestion of Yellen (op. cit.) is that these crises increased the focus of Central Banks around the world on financial stability. However, monetary policy cannot promote financial stability properly. It is, therefore, of great importance that macroprudential policies in terms of "supervision and regulation" (p. 2) should be the primary objectives. Monetary policy is important and "appropriate to ameliorate emerging risks to financial stability" (p. 2). The overall conclusion of Yellen (2014) is that "the pursuit of financial stability is complementary to the goals of price stability and full employment" (p. 3).

Watt (2015) suggests that the inadequate response of the ECB economies to its monetary policy, but also to governments' fiscal policies, and in relation to other advanced countries, explains the weak performance of the Euro Area. However, this could be avoided by introducing a different form of QE, whereby monetary financing of public investment is undertaken. This should be introduced by the ECB by purchasing bonds in the secondary market, issued by the European Investment Bank.[15] The financial resources should be available to national governments to finance projects of productive investment. Such an approach, though, to be effective and efficient, requires a common fiscal EMU policy, and a relevant fiscal authority, along with the relevant monetary policy of the ECB and proper coordination of both policies. The ECB President has recently suggested that what is needed in the Euro Area is coordination of monetary policy with the governments' fiscal policies. European bond markets strengthened after the President's statement, with bond yields falling as prices rose (*Financial Times*, 23 April 2021).

Central Banks after the GFC introduced macroprudential financial stability policies and elevated them to the same level as their inflation target mandate. This is an important development simply because the GFC led to a serious rethink in terms of the Central Banks in relation to macroprudential financial stability policies. Prior to the GFC, Central Banks relied heavily on the theoretical framework of the NCM, whereby achieving price stability would produce financial stability. Clearly, this proposition was completely overrun by the GFC events (Carney, 2020). An interesting question is whether Central Banks should be responsible for financial stability or another Financial Stability Authority. Whatever the answer might be proper coordination of financial stability policies and monetary policy should be properly introduced.

The instruments that can be implemented in terms of macroprudential policies are as follows (Arestis, 2020). Counter-Cyclical Capital Buffer (CCyB). This is the amount of capital, which financial institutions should set aside to avoid breaching their minimum capital requirements; banks and non-bank financial institutions should apply this, and the rest of macroprudential policies.[16] Liquidity Coverage Ratio (PCR), should

[15] The European Investment Bank has been transformed into a 'climate bank'; a very interesting development.

[16] How much capital should financial institutions hold to make them resilient in the case of significant increase in borrower defaults is a relevant question. Mendicino et al.

consist of high-quality assets (cash and short-term government bonds set aside that can be used in a crisis). Debt-to-Income (DTI) ratio, which contains households' indebtedness by imposing a limit to this ratio. Targeted leveraged ratio (total debts /total assets) restricting financial institutions to exceed a minimum ratio. Loan to value ratio (enforcing caps on new loans). On the question of rules versus discretion in applying these macroprudential instruments, discretion is preferable so that policymakers can adjust the use of macroprudential policies as necessary. Coordination with fiscal and monetary policies is vital since such usage of policies is inferior in the stand-alone policies. Especially so since financial stability risks may also occur from Central Banks' QE and FG as suggested by Dragui (2015) at a relevant speech. Should they emerge, Dragui (op. cit.) suggests, macroprudential policies would be suitable to address financial stability risks. A recent example of financial stability policy is China's Central Bank that asked lenders to reduce credit supply in view of the fact that new loan growth reached 16% in the first two months of 2021; thereby causing a significant increase in property prices (*Financial Times*, 05 April 2021).

In terms of the impact of financial stability policies, and macroprudential ones in particular, Carutti et al. (2015) produce relevant evidence in a large sample of countries over the period 200–2013. Carutti et al. (op. cit.) conclude that macroprudential policies are effective in terms of reducing the rate of growth of credit and the rate of growth of house prices (see Bordo, 2017, for further relevant evidence).

Tarullo (2017) examines the case of monetary policy rules, to conclude that "One traditional argument against such a rule has been that there is often too much of importance going on in the economy to be captured by a parsimonious rule and thus produce good policy" (p. 17). The recent GFC clearly supports this proposition. Milton Friedman admitted at a "Lunch with FT: Milton Friedman" (*Financial Times*, 07 June 2003) that "the use of quantity of money as a target has not been a success. I'm not sure that I would, as of today, push it as hard as I once did".

In terms of low interest rates, Tobias (2020) suggests that the current "low-for-long" interest rates contain a potentially serious problem in terms of producing financial instability in the future. Policymakers should

(2021) explore this question to conclude that a capital ratio of around 15% is in fact optimum.

proceed with more stimulus today but should also account for potential problems of financial stability risks in the future.

Jarociński and Maćkowiak (2017) examine the case of fiscal and monetary policies in the Euro Area. They suggest that once fiscal policy is centralized, proper interaction of them would create a much better situation than what has been the case since the GFC. In effect, the argument is that proper interaction of the two policies would produce better results in terms of GDP growth and inflation. Such policy interaction requires the debt to be denominated in a fiat currency that is the euro debt, so that it makes debt non-defaultable.

The chief economist of the OECD, Laurence Boone, argues that countries should continue using fiscal policy (higher government spending and lower taxes) to produce economic recovery and reduce unemployment; a lesson from the GFC is clearly there (as reported in the *Financial Times*, 05 January 2021). Central Banks "are doing whatever it takes to prevent a depression" (*Financial Times*, 04 January 2021). However, since "Monetary policy is a blunt tool", government "spending by contrast can be targeted at those who require it most"; Laurence Boone also warns against the use of austerity after the coronavirus syndrome (*Financial Times*, 04 January 2021)

Following the passing of the USA President's $1.9tn stimulus, 900,000 jobs emerged in March 2021, raising optimism of the stimulus. The USA unemployment rate was reduced to 6% in March 2021 (from 6.2% in February 2021), with the job gains broadly based. 266,000 jobs were created, which was below the 1 m that had been expected; USA businesses have struggled to recruit workers. As reported in *The Economist* (01, 22, and 27, May 2021), a problem with USA unemployment is that despite the 6% unemployment rate, it is difficult to find workers; the number of vacancies is over 8 m, which in fact has never been so high in the past. This is due to generous government benefits, which are higher than the wages in their previous jobs; also, this is in addition to the pandemic syndrome, which induces people to stay at home rather than work (it should be noted that this is also the case in the UK and in Europe). The USA labour department reported that job creation in April reduced significantly (reported in the *Financial Times*, 17 May 2021). This USA situation could potentially increase inflation in view of increases in wages to attract people to work. The annual consumer price index rose in April 2021 at its fastest rate since 2008 (as reported in the *Financial Times*, 17 May 2021). This has produced fears that in view of the "uneven" improvement in the USA labour market, inflationary pressures would emerge. These inflationary worries tend to be dismissed (as stated by Fed officials and reported in the *Financial Times*, 12 May 2021), with the Fed expecting this inflation increase to be transitory. The ECB President has also argued that these concerns about inflation emerging because of "rebounding economic activity" and "supply constraints" are not right because these inflation possibilities are of "temporary nature" (reported in the *Financial Times*, 22 May 2021). Disinflationary pressures could emerge in 2022 in view of possible supply response after the Covid-19 syndrome. Furthermore, if saving rates keep elevated, thereby demand does not increase as supply capacity picks up, government expenditure would not produce inflationary pressures. Disinflationary pressures would be possible under such circumstances.

Not only is the USA stimulus of the current President (a package of $1.9tn) expected to boost the economy, but also other economies like the Euro Area, China, Canada, Mexico and other countries, through increased USA demand for imports from these countries. The OECD (as reported in the *Financial Times*, 22 March 2021) expects a 1% global GDP increase in 2021 because of the USA stimulus. In the case of the USA, and in addition to the fiscal stimulus, creation of bargaining power for the workers, which has weakened since the 1970s, is urgently needed. This would also have positive effects in terms of reducing inequality. Enhancing the power of workers is also relevant to the UK and the EMU.

Not only is the President's $1.9tn fiscal stimulus intended to help the USA to recover from the Covid-19 pandemic, but it also has another goal, namely to reduce inequality. Inequality in the USA and elsewhere has increased since the GFC and the Covid-19 periods. An example of the $1.9tn fiscal stimulus in terms of reducing inequality is that it proposes resources for the poor, which should help to reduce inequality in the USA. In addition to the $1.9tn fiscal stimulus, the current President has suggested another $3tn in terms of infrastructure, healthcare, climate issues, energy and education, partially funded by tax increases on corporations and wealthy. The President has proposed $2.3tn to be provided for infrastructure, along with raising the corporate tax rate from 21 to 28%. Subsequently, the President proposed a global tax rate of at least 15%, down from his original proposal of 21%; the G20 will discuss this issue in July 2021, which would be a good opportunity to reform it so that countries would agree on it).[17] The G7 is very close to agreeing with the President's global corporate tax proposal, which would help the G20 to reach a relevant agreement. On the 28th of May 2021, the USA President unveiled his infrastructure plan to $6tn budget proposal.

All this should help increase the supply side of the economy, which is necessary in view of signs that insufficient supply is emerging, especially so in view of the business struggling to recruit workers as mentioned above. Increase in the supply helps inflation not to reach high levels, and, in addition, helps to increase productivity. More resources for public investment add to productive capacity thereby producing growth and aggregate demand. Increasing aggregate demand enhances productivity. The $2.3tn infrastructure plan (such spending in the USA was at the lowest level over the last six decades) is proposed to be funded through corporation tax increases, which is not acceptable to the large USA business, which suggests it is too high, raising tensions that are bound to complicate the prospects of the plan.[18] In fact, the 28% corporate tax rate is not excessively high in view of it being 35% before the previous President reduced it in 2017. In the UK the Chancellor of the Exchequer in the March 2021 budget he announced a gradual rise of the corporate taxation to 25% from the current 19%, and from 2023.

[17] The USA Treasury Secretary, Janet Yellen, suggested in her speech to the Chicago Council on Global Affairs on 05 April 2021, the G20 and other countries should join the USA to set a global minimum corporate tax. This in her view would avoid profit shifting across countries to avoid taxes. Companies would pay taxes where their revenues are earned, not where their profits would be shifted—normally to tax heavens. In addition, this is needed to create a "stable tax system that raise sufficient revenue to invest in essential public goods and respond to crises" (as reported in the *Financial Times*, 06 April 2021). Janet Yellen also suggested on 07 April 2021 (as reported in the *Financial Times* on 08 April 2021) that the $2tn plan would produce 1.6% increase to GDP by 2024, alongside roughly $2.5tn in corporate tax increase. According to the Treasury Secretary, these are mutually beneficial policies. European countries have welcomed the suggestion of a global corporate tax, but they disagree over details of the proposal, especially over the minimum corporate tax (*Financial Times*, 15 April 2021); The UK's Chancellor of the Exchequer is holding back support of the global corporate tax, arguing that it should be part of a broader package of a fairer system (reported in the *Financial Times*, 17 May 2021). Janet Yellen (on May 2021), urged Congress to accept higher taxes to pay for the President's plan of spending. Janet Yellen also suggested that if inflation became an issue, tools are available to address it (reported in the *Financial Times*, 03 May 2021).

[18] The USA President has suggested recently (mid-April 2021) to reduce his infrastructure bill by 60bn in an attempt to get his infrastructure bill through the Congress.

It should be noted that public infrastructure investment has its impact over the medium to long run, simply because it needs time to have its effects. However, under current circumstances of spare capacity, low levels of income, near zero interest rates, and high uncertainty, infrastructure can be expansionary even in the short run.

An interesting development has recently occurred in the USA where the Treasury Secretary and the Chair of the Fed appear together before the Congress. This is an important development, which raises the question of whether the Treasury and the Fed would continue so and coordinate their policies. In addition, whether similar governments and Central Banks would copy the USA and proceed along similar lines. It should also be noted that the recent USA policy mix, and in other countries, whereby fiscal policy is employed in addition to unconventional monetary policies, is a definite improvement in relation to the single monetary policy initiated after the GFC. The USA President's fiscal stimulus has had an accelerating recovery in the first quarter of 2021, close to the pre-pandemic level. The GDP grew by 6.4% in comparison to the first quarter of 2020 (1% lower). Alternatively, 1.6% compared to the 4th quarter GDP increase. The Fed has kept its QE at $120bn per month, despite its acknowledgement of the recent expansion of economic activity (as reported in the *Financial Times*, 30 April 2021).

In terms of public debt, the IMF's latest fiscal monitor database, published on 28 January 2021 (as reported in the *Financial Times*, 29 January 2021), suggests that fiscal authorities should not worry about high levels of public debt, especially under the very low interest rates, and should not view it as priority. Fiscal policy should be employed to stimulate and restore economic growth and reduce unemployment. Fiscal policy has the ability to respond rapidly when implemented properly.[19] The IMF's relaxed approach to high public debt levels emerged in view of the levels of interest rates being close to zero. There is also the view that fiscal policy may very well increase the level of economic activity but there are serious dangers of inflation. When inflation occurs, the Central Bank would increase its rate of interest, which would harm the level

[19] Canzoneri et al. (2016) examine fiscal multipliers in the case of recessions, employing a model based on Cardia and Woodford (2009, 2010), to show that fiscal multipliers are state dependent, and during recessions their values exceed two, while during booms they are less than one.

of economic activity and increase unemployment; this serious inflation-driven rise in interest rates, could potentially produce a crisis. However, if fiscal policy were designed so that more capacity is generated to help create higher future growth, inflation would be prevented. It should also be noted that reducing income and wealth inequality should be a prime objective of fiscal policy. The sizable stimulus of the current USA President, along with loose monetary policy, the accumulated savings of consumers and falling unemployment as well as strong subsequent recovery, raises the issue of producing inflationary pressures. The recent USA monetary policy change, with an indication of tolerating the inflation rate to rise above the Fed's inflation target should help. In any case, the state of the USA labour market suggests that inflation should be kept subdued. The USA labour market is about 8.4 m jobs short of what it was in February 2021 (*The Economist*, 17 April 2021). However, attention to high inflation rates is needed so that the Fed, and of course other Central Banks, act timely along with proper coordination with fiscal policy and financial stability. It should be noted, though, that inflation reduces the value of debt and thereby easing the debt burden. This, however, transfers wealth from savers to borrowers.

Shadow banking could be one of the main causes of another financial crisis, as it was in the case of the GFC. Shadow banking institutions, are financial institutions, which undertake bank-like activities, but are not regulated as the normal banks. Actually, some regulations have been undertaken, such as requiring shadow banks to hold more liquid assets, limits on leverages, and oversight of the shadow banking system by the Central Banks; but not sufficiently to avoid contributing to another similar crisis to GFC.[20] Becard and Gauthier (2021), suggest, "These institutions handle vast amounts of debt and serve as an additional source of funding for commercial banks. But they typically exhibit a volatile and procyclical behaviour" (p. 2).

Increases in inequality influence "secular stagnation" (Bank of England, 2015, p. 30) and increase risks to financial stability (Rajan, 2010). This is all due to widening access to credit extensively. A good example of such a case had occurred prior to the Great Depression, and prior to the GFC (Kumhof et al., 2013). In both of these cases "a link

[20] In the case of the UK's Financial Conduct Authority (FCA), it has been criticized for failing to regulate properly the sub-prime lending (as reported in the Observer, 04 April 2021).

is established between rising inequality and demand for credit, and ultimately a financial crisis" (Bank of England, 2015, p. 30; see, also, Arestis, 2016).

A further point is whether the Credit Rating Agents have been required to disclose more of their methodology to investors. This is to avoid similar contribution to future crises of the GFC type, whereby they warn investors properly of the true risks of the securities, they rate. Relevant proposals have been suggested, for example, replacing private credit ratings with those of a public agency (Rutledge and Litan, 2014). Still no relevant proposals have been implemented.

Friedman (2015) poses an interesting question whether the GFC has changed permanently monetary theory and policy. Friedman (op. cit.) suggests that relevant monetary policy changes, especially the unconventional ones, have had profound implications for monetary theory and policy. The reason suggested for the success of these new unconventional monetary policies is twofold. "First these actions have been reasonably successful on their own terms. As intended, when used they have lowered long-term interest rates relative to short-term rates and lowered interest rates on more-risky compared with less-risky obligations. And second ... their introduction fills a conceptual vacuum that has stood at the heart of monetary policy analysis for the better part of the last century ever since the working out of modern central banking" (Friedman, 2015, pp. 7–8). FG, however, has not been so successful; nor has the traditional "quantity theory" in view of the enormous increase in the supply of money without inflation at all! A further change of monetary policy is that both instruments of Central Banks, the short-term interest rate, and their quantity of assets and liabilities, are two independent instruments. Therefore, "monetary policy has not one independent instrument but two" (Friedman, 2015, p. 18).

Tressel and Zhang (2016), using the ECB Bank Lending Survey (BLS), assess theoretically and empirically the extent to which macroprudential policies in terms of "credit growth and house price appreciation in mortgage markets" are effective. Their conclusion is that most effective is the "cost of bank capital", which reduces the mortgage credit growth. Targeting the Loan-to-Value (LTV) ratio[21] is also effective, especially

[21] The LTV ratio is a policy tool that relates to the housing market. It is a lending risk assessment that financial institutions and other lenders consider before approving a mortgage. Higher LTV implies higher risk loans.

when monetary policy is very loose; this is so in view of macroprudential policies' interactions with monetary policy.

An alternative to QE and FG, and important monetary policy instrument, is what is called the "yield curve control"; namely, long-term interest rate targeting that aims to achieve a specific level along the yield curve. The Central Bank, therefore, should decide on the relevant target of bond yields that is consistent with price stability, and then proceed with bond purchasing as necessary to achieve its target (Barwell, 2020). "Yield curve control" is different from QE. The latter rely on purchasing mostly government debt to lower long-term interest rates, while the "yield curve control" caps yields at specific levels of the "yield curve". This monetary policy instrument means that the Central Bank should directly set the long-term interest rate. The Central Bank of Japan introduced this type of policy instrument in 2016, and continues to implement it until now, aiming at keeping the 10-year Japanese government bond interest rate near zero. The Reserve Bank of Australia, and in March 2020, started targeting the 3-year interest rates. Such policy aims to achieve directly what QE can achieve indirectly; "yield curve control" provides lower long-term interest rates directly, thereby encouraging productive investments to be undertaken, and enabling the Central Bank to achieve its inflation target, more promptly than QE and FG. In terms of economic policies, both demand-side and supply-side policies are necessary. This is simply because although demand-side policies enhance the level of economic activity, if insufficient supply is evident, inflation emerges. In addition, Bailey (2021) suggested, it is the case that insufficient spare capacity could have an important impact on inflation; therefore, both demand and supply are important on this issue.

Sustained rates of business investment increase the productive capacity of the economy. Consequently, rapid recovery of business investment should be the priority of policymakers in their attempt to enhance the rate of economic growth (see, also, Smith, 2022). Investment in new technology and manufacturing, infrastructure,[22] investment in human infrastructure, investment in science and skills, all of them should be

[22] In the UK, a new Infrastructure Bank is planned; the UK Treasury announced it, and confirmed by the Chancellor on his budget day (03 March 2021). This new bank is to be launched in spring 2021 to support private investment projects. The Chancellor on 03 March 2021 declared that the Bank of England should have another objective, namely environmental sustainability.

an important dimension of fiscal stimulus. In addition, green public investment is very important to sustainable development (Verselli, 2017, pp. 203–210).[23] This combination ensures the economy's sustainability, not just cyclical stability. This requirement is helped greatly by proper coordination of fiscal, financial stability and monetary policies. In fact, recent cooperation between the Treasury and the Central Bank in some countries in view of the Covid-19 has shown that such cooperation has powerful effects on the real economy and financial markets. Continuing so and in terms of close coordination with financial stability, should have very positive effects after the current pandemic. A recent example of such coordination is the Bank of England and the HM Treasury announcement to create jointly a Central Bank Digital Currency (CBDC) Taskforce, which will coordinate the potential creation of the proposed CBDC).[24] The Bank of England would issue the CBDC in a new form of money to be used by households and businesses. Both the Deputy Governor for Financial Stability and the HM Treasury Director General will chair the Taskforce. Clearly, financial stability will also be taken on board in this coordination. A further relevant UK example is the November 2020

[23] A report published on 24 March 2021 (Network for Greening the Financial System; a group of 89 Central Banks and financial supervisors formed in relation to the Paris climate goals), advises central banks how to avoid climate change risks, and support their national governments, without influencing their monetary policies. The main proposal is for Central Banks to offer a lower rate when financing banks, which meet climate-related lending criteria (*Financial Times*, 25 March 2021). 'Green Quantitative Easing', and 'Green Targeted Lending Operations', are additional policies to enable energy transition. Clearly, Central Banks should not ignore climate changes, which affect economies profoundly. Indeed, central banks should be involved with climate change. They can, and should, accelerate reductions in carbon emissions, change the cost of capital, use stress tests, to address climate risks in the financial sector, and apply relevant macroeconomic policies to avoid financial instability. It should also be noted that the virtual climate summit meeting (with 40 world leaders) organized by the USA President on the 22nd and 23rd of April 2021, and dealt with carbon emissions, pledged, "to cut emissions by at least 50 percent by the end of this decade" (*Financial Times*, 24 April 2021). The UK pledged to catch its emissions by 78% and by 2035. The summit, though, ended with very few funding promises for the poor countries. However, the Bank of England, and other Central Banks, are examining how they can orderly contribute to the emergence of a carbon-neutral economy (Ramsden, 2021).

[24] As reported in *The Economist* (08 May 2021), in the USA the Fed is already considering the possibility of a CBDC. Also reported is that the CBDC is hoped to be introduced by 2025 by the ECB. Moreover, such conducive economic policies can significantly support progressive social, environmental, and economic development strategies and policies.

creation by the Treasury, the Bank of England and the Financial Conduct Authority of a Working Group on Productive Finance. The membership of the Group includes market participants and interested bodies. The focus of the Group is on supporting the UK's economic growth by promoting more investment and infrastructure. This is an important development in view of when government and business work together as part of a national economic strategy, it helps to increase investment, especially infrastructure, which is of paramount importance.

Our argument on the importance of coordination of fiscal policy and monetary policy and financial stability is also supported by empirical evidence. Ryan-Collins and van Lerven (2018) examine fiscal and monetary policy coordination over the period 1930s–1970s, and conclude that there is a great deal of evidence that suggests clearly that such coordination was very important in terms of the impact of the coordinated policies on the economy. Eggertsson (2006) also provides relevant empirical support and concludes that the multipliers increase substantially when monetary and fiscal policies are coordinated. We would add of course that coordination of fiscal, monetary and financial stability policies becomes paramount, especially for the post-neoliberal world.

4 Summary and Conclusions

The focus of this contribution is on the relevant macroeconomic policies employed during the GFC and Covid-19, and the required economic policies post them. The relevant economic policies suggested are fiscal, monetary and financial stability. The latter had not been introduced prior to the GFC, but even since then it has not been properly accounted for and introduced. A relevant argument of this contribution is that financial stability should be the primary objective of the Central Banks' policies.

Fiscal, monetary and financial stability policies have been discussed in this contribution. The most important discussion and proposal that follows, is that these policies need to be properly coordinated if sustainable economic activity is to be achieved. It is also suggested that these policies should include in their objectives inequality. Such coordination should generate a path for resilient and sustainable growth, enhance productivity and reduce inequality. In terms of inequality, a relevant recent development has emerged. This is the strength of trade union membership, which since 1985 and across OECD countries has been halved on average (*Financial Times*, 08 March 2021). The good news now is that

Covid-19 has prompted significantly more workers to join trade unions. The latter have the best opportunity to become important institutions in the economy, which should help reduce inequality. Interestingly enough, Bailey (2021) suggests that further policy changes, both demand and supply ones, are needed for the post-pandemic era. Our suggestion of proper coordination of fiscal, monetary and financial stability policies is the best way forward. Governments and Central Banks around the world should move towards such coordination of policies. Fiscal and monetary authorities should do their best to keep such coordination safe, resilient and transparent.

References

Acocella, N., Di Bartolomeo, G. and Hughes Hallet, A. (2016), *Macroeconomic Paradigms and Economic Policy: From the Great Depression to the Great Recession*, Cambridge: Cambridge University Press.

Angeriz, A. and Arestis, P. (2008), "Assessing Inflation Targeting Through Intervention Analysis", *Oxford Economic Papers*, 60(2), pp. 293–317.

Arestis, P. (2016), "Main and Contributory Causes of the Recent Financial Crisis and Economic Policy Implications", in P. Arestis and M. C. Sawyer (eds.), *Emerging Economies During and After the Great Recession*, Annual Edition of *International Papers in Political Economy*, (Palgrave Macmillan, 2016)

Arestis, P. (2019a), "The Past and Future of the Euro", *Revista de Economia Critica*, Vol. 27, No. 2, pp. 6–15.

Arestis, P. (2019b), "Critique of the New Consensus Macroeconomics and Propose a More Keynesian Macroeconomic Model", in P. Arestis and M. C. Sawyer (eds.), *Frontiers of Heterodox Macroeconomics*, Annual Edition of *International Papers in Political Economy*, London: Palgrave Macmillan.

Arestis, P. (2020), "Financial Stability: Still Unsettled for the Future", in P. Arestis and M.C. Sawyer (eds.), *Economic Policies for a Post Neo-Liberal World* , Annual Edition of *International Papers in Political Economy*, London: Palgrave Macmillan.

Arestis, P. and Sawyer, M. (2011), "The Design Faults of the Economic and Monetary Union", *Journal of Contemporary European Studies*, 19(1), pp. 19–30.

Arestis, P. and Sawyer, M. (2019), "Critique of the New Consensus Macroeconomics and Propose a More Keynesian Macroeconomic Model", in P. Arestis and M. Sawyer (eds.), *Frontiers of Heterodox Macroeconomics*, Annual Edition of International Papers in Political Economy, London: Palgrave Macmillan.

Arestis, P. and González Martinez, A.R. (2015), "The Absence of Environmental Issues in the New Consensus Macroeconomics is only One of Numerous Criticisms". In P. Arestis and M. Sawyer (eds.), *Finance and the Macroeconomics of Environmental Policies*, Annual Edition of *International Papers in Political Economy*, London: Palgrave Macmillan.

Bailey, A. (2021), "Getting over Covid", *Speech given at the Resolution Foundation*, 8 March 2021. Available at: www.bankofengland.co.uk/news/speeches

Bailey, A., Bridges, J., Harrison, R., Jones, J. and Mankodi, A. (2020), "The Central Bank Balance Sheet as a Policy Tool: Past, Present and Future", *Bank of England Staff Working Paper No 899*, Bank of England, December 2020. Available at: www.bankofengland.co.uk/working-paper/staff-working-papers

Balima, H.W., Kilama, E.G. and Tapsoba, R. (2017), "Settling the Inflation Targeting Debate: Lights from a Meta-Regression Analysis", *IMF Working Paper WP/17/213*, September 2017.

Bank of England (2015), "On Bank Research Agenda", *Discussion Paper*, Bank of England.

Barwell, R. (2020), "Bank of England Should Switch Strategy on QE", *Financial Times*, 10 December 2020.

Becard, Y. and Gauthier, D. (2021), "Banks, Shadow Banks, and Business Cycles", *Bank of England Staff Working Paper No. 907*, February 2021

Blanchard, O. (2018), "On the Future of Macroeconomic Models", *Oxford Review of Economic Policy,* **34**(1–2), pp. 43–54.

Blinder, A., Ehrmann, M. de Haan, J. and Jansen, D.-J. (2017), "What Will Monetary Policy Look Like After the Crisis", *European Central Bank Research Bulletin*, No. 39, 18 October 2017.

Bordo, M.D. (2017), "An Historical Perspective on Financial Stability and Monetary Policy Regimes: A Case for Caution in Central Banks Current Obsession with Financial Stability", *Norges Bank Working Paper 5/2018*, November 2017.

Canzoneri, M., Collard, F., Dellas, H. and Diba, B. 2016), Fiscal Multipliers in Recessions", *The Economic Journal*, 126(590), pp. 75–108.

Carney, M. (2020), *Building a Better World for All*, London: William Collins.

Carutti, E., Claessens, S. and Laeven, L. (2015), "The Use and Effectiveness of Macroprudential Policies: New Evidence", *IMF Working Paper WP/15/61*, International Monetary Fund, March 2015.

Colciago, A., Samarina, A. and de Haan, J. (2018), "Central Bank Policies and Income and Wealth Inequality: A Survey", *DNB Working Paper No. 594*, May 2018, De Nederlandsche Bank, Amsterdam, The Netherlands.

Curdia, V. and Woodford, M. (2009), "Credit Functions and Optimal Monetary Policy", *Working Paper No. 278*, Board of International Settlements.

Curdia, V. and Woodford, M. (2010), "Credit Spreads and Monetary Policy", *Journal of Money, Credit and Banking*, 42(S1), pp. 3–35.

Dixit, A. and Lambertini, L. (2001), "Monetary-Fiscal Policy Interactions and Commitment versus Discretion in a Monetary Union", *European Economic Review*, 45, pp. 977–997.

Dixit, A. and Lambertini, L. (2003), "Symbiosis of Monetary and Fiscal Policies in a Monetary Union", *Journal of International Economics*, 60, pp. 235–247.

Draqui, M. (2015), "The ECB and its Watchers XVI Conference", *Speech Delivered at Frankfurt Main*, 11 March 2015.

Driffill, J. (2015), "Unconventional Monetary Policy in the Euro Zone", *Working Paper ISSN 2183–1815*, Lisbon School of Economics & Management, Department of Economics.

Eggertsson, G. B. (2006), "Fiscal Multipliers and Policy Coordination", *Federal Reserve Bank of New York Staff Reports, No. 241*, F New York: Federal Reserve Bank of New York.

Ehnts, D. and Paetz, M.(2021), "COVID-19 and its economic consequences for the Euro Area", *Eurasian Economic Review.* Available at: https://doi.org/https://doi.org/10.1007/s40822-020-00159-w

Elbourne, A., Ji, K. and Duijndam, S. (2018), "The Effects of Unconventional Monetary Policy in the Euro Area", *CPB Discussion Paper 371*, CPB Netherlands Bureau for Economic Policy Analysis.

Friedman, B.M. (2015), "Has the Financial Crisis Permanently Changed the Practice of Monetary Policy? Has it Changed the Theory of Monetary Policy?", *The Manchester School*, 5(11), Supplement 2015.

Goodhart, C. (2012), "Monetary Policy and Public Debt", *Banque de France Financial Stability Review*, 16, pp. 123–130, April 2012.

Goodhart, C. and Ashworth, J. (2012), "QE: A Successful Start May Be Running Into Diminishing Returns", *Oxford Review of Economic Policy*, 28(4), pp. 640–670.

Haldane, A. (2021), "Inflation: A Tiger by the Tail?", *Speech Given Online*, 26 February 2021. Available at: https://www.bankofengland.co.uk/news/speeches

Haldane, A., Roberts-Sklar, M., Wieladok, T. and Young, C. (2016), "QE: The Story so Far", *Bank of England Working Paper 624*, Bank of England, October 2016.

IMF (2015), "Negative Interest Rates in Europe: A Glance at Their Causes and Implications", in *Global Economic Prospects*, Chapter 1, June 2015.

IMF (2017), "Negative Interest Rate Policies – Initial Experiences and Assessments", *Staff Report*, 13 March 2017. Available at: http://www.imf.org/external/pp/ppindex.aspx

Jarociński, M. and Maćkowiak, B. (2017), "Monetary Fiscal Interactions and the Euro Area Malaise", *NBER Working Paper 23746*, August 2017. Available at: http://www.nber.org/papers/w23746

King, M. (2016), *The End of Alchemy: Money, Banking and the Future of the Global Economy*. London: Little Brown.

King, T. B. (2019), "Expectation and Duration at the Effective Lower Bound", *Journal of Financial Economics*, 134, pp. 736–60.

Kumhof, M., Ranciere, R. and Winant, P. (2013), "Inequality, Leverage and Crises: The Case of Endogenous Default", *IMF Working Paper, 13/249*, International Monetary Fund.

Leeper, E. (1993), "The Policy Tango; Towards a Holistic View of Monetary and Fiscal Effects", *Economic Review*, Federal Reserve Bank of Atlanta, 78(4), pp. 1–27.

McCausland, W.D. and Theodossiou, I. (2016), "The Consequences of Fiscal Stimulus on Public Debt: A Historical Perspective", *Cambridge Journal of Economics*, 40(4), pp. 1103–1116.

Mendicino, C., Nikolov, K., Rubio-Remirez, J. Suarez, J. and Supera, D. (2021), "How Much Capital Should Banks Hold?", *Research Bulletin No. 80*, European Central Bank, 27 January 2021. Available at: https://www.ecb.europa.eu/pub/economic-research/resbull/2021

Muellbauer, J. (2020), "Implications of Household-level Evidence for Policy Models: The Case of Macro-Financial Linkages', *Oxford Review of Economic Policy*, **36**(3), pp. 510–55.

Orphanides, A. (2017), "The Fiscal-Monetary Policy Mix in the Euro Area: Challenges at the Zero Lower Bound", *MIT Sloan School Working Paper 5197–17*, 8 May 2017. Available at: http://creativecomments.org/licenses/by-nc/

Rajan, R. (2010), *Fault Lines: How Hidden Fractures Still Threaten the World Economy*, Princeton University Press, Princeton, USA.

Ramsden, D. (2021), "QE as an Economic Policy Tool—What does it do and how should we Use it?". Speech given at the Peter Sinclair Town Hall, University of Birmingham, 17th February 2021.

Rutledge, A. and Litan, R.E. (2014), "A Real Fix for Credit Ratings", *Economic Studies at Brookings*, June 2014.

Ryan-Collins, J. and van Lerven, F. (2018), "Bringing the Helicopter to Ground: A Historical Review of Fiscal-Monetary Coordination to Support Economic Growth in the 20th Century", UCL Institute for Innovation and Public Purpose (IIPP) Working Paper Series 2018–08. Available at: https://www.ucl.ac.uk/bartlett/public-purpose/wp2018-08

Sawyer, M, (2013), "Alternative Economic Policies for the Economic and Monetary Union", *Contributions to Political Economy*, 32(1), pp. 11–27.

Sinclair, P. and Allen, W.A. (2017), "Monetary Policy Normals, Future and Past", National Institute Economic Review, No. 241, pp. R5–R12.

Skidelsky, R. (2017), "How Would Keynes have Analysed the Great Recession of 2008 and 2009?", ch. 9 in T. Congdon, *Money in the Great Recession: Did a Crash in Money Growth Cause the Global Slump?*, Edward Elgar Publishers, UK.

Smith, J. (2021), "The Macroeconomic Policy Outlook", *Macroeconomic Policy Unit,* Quarterly Briefing Q1 2021. London: Resolution Foundation.

Stiglitz, J.E. (2010), *Economics of the Public Sector*, New York: The New Press.

Stiglitz, J.E. (2012), "Macroeconomics, Monetary Policy, and the Crisis", in O.J. Blanchard, D. Romer, M. Spence and J.E. Stiglitz (eds.), *In the Wake of the Crisis: Leading Economists Reassess Economic Policy.* Cambridge (MA): MIT Press.

Stiglitz, J.E. (2015), "New Theoretical Perspectives on the Distribution of Income and Wealth Among Individuals: Part IV: Land and Credit", *NBER Working Paper 21192*, Cambridge (MA): National Bureau for Economic Research.

Tarullo, D.K. (2017), "Monetary Policy without a Working Theory of Inflation", Hutchins Center Working Paper 33, October 2017.

Tenreyro, S. (2021), "Response to the Covid-19 Pandemic: UK and US Experiences". *Speech Given at the Federal Reserve Bank of San Francisco Macroeconomics and Monetary Policy*, Spring Conference, 26 March 2021.

Tobias, A. (2020), "What to do When Low-for-Long Interest Rates are Lower and for Longer", *IMF Blog*, 14 December 2020.

Tressel, T. and Zhang, Y.S. (2016), "Effectiveness and Channels of Macroprudential Policies: Lessons from the Euro Area", *IMF Working Paper WP/16/4*, Strategy, Policy, and Review Department, IMF, January 2016.

Verselli, A. (2017), *Crisis and Sustainability: The Delusion of Free Markets*, London: Palgrave Macmillan.

Vines, D. and Wills, S. (2020), "The Rebuilding Macroeconomic Theory Project Part II: Multiple Equilibria, Toy Models, and Policy Models in a New Macroeconomic Paradigm", *Oxford Review of Economic Policy*, 36(2), pp. 427–497.

Wren-Lewis, S. (2020), "Mapping Types of Macromodel to Types of Macro User", *Oxford Review of Economic Policy,* **36**(3), pp. 498–509.

Yellen, J.L. (2014), "Monetary Policy and Financial Stability", T*he 2014 Michel Camdessus Central Banking Lecture*, International Monetary Fund, Washington D.C., 2 July 2014.

Watt, A. (2015), "Quantitative Easing with Bite: A Proposal for Conditional Overt Monetary Financing of Public Investment", *IMK Working Paper No. 148*, March 2015, IMK Macroeconomic Institute. Available at: http://www.boeckler.de/cmk5016.htm

Monetary Policy Support for a Green New Deal

Sergio Rossi

1 Introduction[1]

All around the world, and particularly in so-called "advanced economies", the covid-19 pandemic has revealed the flaws of neoliberalism, notably in respect of (a lack of) State intervention and proper regulation of economic activity (for a thorough explanation see notably Karagiannis and King 2019). As regards monetary policy, it is plain that central banks should not limit their focus to price stability (on the market for produced goods and services), but must first and foremost act to guarantee financial stability for the economic system as a whole—which has become a major issue after the outbreak of the global financial crisis in 2008 (see Carney 2015; Danmarks Nationalbank 2019, but also Rossi 2015). Now,

[1] The author is grateful to the editors of this volume for their invitation to contribute a chapter as well as for their comments on an earlier version of it. He is also indebted to Carryl Oberson and Maurizio Solari for their suggestions in this regard. The usual disclaimer applies.

S. Rossi (✉)
Department of Economics, University of Fribourg, Fribourg, Switzerland
e-mail: sergio.rossi@unifr.ch

N. Karagiannis and J. E. King (eds.), *Visions and Strategies for a Sustainable Economy*, Global Institute for Sustainable Prosperity,
https://doi.org/10.1007/978-3-031-06493-7_5

owing to the urgent needs to act against the huge damages caused by climate change all over the world, monetary policy should also contribute to a collective effort in supporting and carrying out an ecological transition that favours economic development for the common good, thereby satisfying people's needs (see Bolton et al. 2020). This is the reason why the public sector must promote a green new deal with the monetary and financial support of central banks.

In the next section we explain that the general government sector should issue "green bonds" to finance the ecological transition in a sustainable way. Central banks should buy these bonds on the primary market and introduce them in the list of eligible assets to be used as collateral by any banks in need of liquidity on the interbank market. The third section then argues that central banks should steer their policy rates of interest with a view to support and enhance the ecological transition. This amounts to reducing the interest rate for those loans that support such a transition, particularly as regards firms' investments. Central banks will thereby contribute to make the world a better place to live in, both as regards climate conditions and financial stability. In such a framework, firms' business will benefit from central banks' decisions, as far as the former contributes to achieve the policy targets of the latter. The last section concludes by pointing out that social cohesion will be enhanced by monetary policy strategies favouring a green new deal. If so, the world will become a better place to live in, with a sustainable economy over the long run.

2 Green Government Bonds for a Sustainable Economy

Greening the global economy has become a priority and features in a number of policy agenda of so-called "advanced economies" (see Campiglio 2016; Krogstrup and Oman 2019). At the same time, the global economy is in a strong recession, as a result of the covid-19 pandemic and the ensuing lockdown of a variety of economic activities all around the world. The so-called golden rule of public finance should be considered in this regard, in order to promote economic growth and make it much more favourable for the environment than this is the case to date. As is well known, the golden rule of public finance considers that public investment must be financed through public debt, because it benefits several generations of taxpayers (see Eisner 1984, 1992; Dafflon

1996). Indeed, it would be ethically wrong to ask the current generation of taxpayers to finance public investment entirely, that is, with a balanced budget for the general government sector, ignoring that also future generations of taxpayers are going to benefit from such an investment. This is so much so considering that fiscal policy, as argued by Lerner (1944), should aim at full employment through a countercyclical stance: have a fiscal deficit when there is a lack of demand on the market for produced goods and services, and a fiscal surplus in the contrary case (see King 2023). Such a "functional finance" principle for fiscal policy should be applied through the capital account of the general government sector, that is, the balance sheet recording all public investments and the identically equivalent government debt necessary to finance them.

In order to make sure that each generation of taxpayers contributes to financing public investment as much as it benefits from the latter, the pay-as-you-use principle should be put into practice through public debt and taxation. Each generation of taxpayers should therefore pay taxes for an amount that corresponds to (*i*) the current expenditures of the public sector and (*ii*) amortization and interest payment for the accumulated public debt during the relevant fiscal year. The amount of this debt service (that is, amortization and interest payment) must therefore be included in the public sector's current account, to be balanced yearly with fiscal revenue (as current expenditures of the public sector benefit the current generation of taxpayers only), while public investment has to feature in the public sector's capital account, which should be financed entirely through the issuance of government bonds—to wit, giving rise to a public debt that corresponds to the assets thereby produced for the common good of current and future generations of taxpayers.

Empirical evidence as well as post-Keynesian economics clearly show that a variety of public investments have a positive effect on economic growth, owing in particular to so-called "crowding-in" effects—that is, the stimulus of private investments that comes as a result of public investments, such as in the case of transportation, health services or education (see for instance Sawyer 2016). Now, in light of both the covid-19 pandemic and the climate crisis that are affecting the global economy at the time of writing, there is an urgent need for greening economic growth. Public investment in infrastructure, such as health care, smart power grids and telecommunication networks, is therefore crucial to greening economic growth across the world economy. Financing this kind of projects requires the issuance of green government bonds, because it is only through public debt that the pay-as-you-use principle can be respected in this subject matter as explained above.

These green government bonds are likely to become AAA assets, since they kick-start economic growth and make the economic system much more sustainable than it is the case actually. They may thus become secure investment opportunities for institutional investors such as pension funds. They can thereby also become an important instrument for the conduct of monetary policy, since central banks will include them in their lists of eligible assets that banks and non-bank financial institutions may use as collateral to borrow across the interbank market or from the relevant central bank.

3 Interest-Rate Policy for an Ecological Transition

Central banks can contribute to an ecological transition for a sustainable economy in at least two ways (for some alternative proposals see Volz 2017; Campiglio et al. 2018; Dafermos et al. 2018). As explained in the previous section, the first way is to buy green bonds on the primary market, where the general government sector issues them in order for it to finance public investment in environment-friendly economic activities. In doing so, central banks should not put any pressure on the general government sector as regards the amount of public debt also with respect to the country's gross domestic product: independently of the balance of government accounts, indeed, the national central bank should support any public spending for a green new deal by purchasing government bonds on the primary market and without any limit. As a matter of fact, central banks are not profit-oriented financial institutions like banks, which set their lending rates of interest in order to maximize their profits over the short run (without any consideration for ecologically sustainable economic activities). Hence, the central bank can (be willing to) fix a lower rate of interest that the general government sector will have to pay when borrowing from it than this is the case with commercial banks or non-bank financial institutions that would purchase government bonds on the primary market.

The second way by which central banks can support an ecological transition is to steer their policy rates of interest to induce firms as well as households to borrow from commercial banks to finance an ecological transition to a sustainable economy. This could occur through a variety of so-called "transmission channels" stemming from the interest rates decided by the central bank, which thereby influences a number of market rates of interest, thus affecting different magnitudes like exchange rates, foreign trade, consumption and investment—all of which have sooner or later an impact on employment and price levels. Let us expand on this by looking at the monetary policy transmission mechanism with regard to the ecological transition at stake.

This transmission mechanism stems from the policy rates of interest that central banks set up to affect the whole structure of market rates of interest and thereby the (targeted) rate of inflation eventually. Generally speaking, the central bank determines the interest rate that banks will have to pay when they borrow from it the amounts they need to pay their debt on the interbank market (which is the market where banks lend to and borrow from each other, to carry out their transactions across the financial market). This policy rate of interest is in a position to influence interbank market rates of interest, because a bank may choose to borrow from other banks or from the central bank—depending on the level of policy and interbank market rates of interest. Hence, when the central bank wants to reduce interest-rate levels across the relevant economy, it lowers its own policy rate, which induces banks to borrow more from it, thereby reducing interbank lending and as a result also the interbank market rates of interest. This is likely to affect thereby aggregate demand, supporting economic growth and the employment level, since banks are thus induced to reduce their lending rates of interest for both firms and households—both of which could then be more inclined to increase their spending through a higher volume of bank credit.

In this regard, the central bank may decide to lower its policy rates of interest more for those banks whose lending decisions support the ecological transition by opening credit lines with favourable conditions for those firms and households whose activities allow greening the economic system. It may also reduce less, or even increase, the policy rates of interest for those banks whose lending policy aggravates the environmental damages of economic activity (see Svartzman et al. 2019). Such an incentive-based interest-rate policy is in a position to have an impact on agents' behaviour that will contribute to greening the economic system in a financially sound and ecologically sustainable way (see D'Orazio and Popoyan 2019).

Something similar could also occur as regards the asset portfolio of banks: those banks whose portfolio is made of financial assets linked to environment-friendly economic activities may borrow from the central bank at more favourable rates of interest than what other banks may do. The former banks might also benefit from a higher interest rate on their deposits at the central bank than occurs for the latter banks—whose central bank deposits might be remunerated with a lower rate of interest (becoming a zero or negative rate of interest, if the relevant bank's portfolio contains a higher share of financial assets that are linked to damaging economic activities for the environment).

These impacts of interest-rate policy on greening the economic system can be enhanced by other "sub-channels" through which the transmission mechanism operates. Indeed, a so-called "balance-sheet channel" can impact firms' balance sheets when the policy rate of interest decided by the central bank to induce an appropriate ecological transition has an impact on the composition of these balance sheets. Such an impact can occur through both the amount of the firms' cash flow and the type of financial assets in their portfolio (made of "brown" or "green" assets). Firms whose cash flow much depends on greening the economic system may benefit from more favourable rates of interest decided by commercial banks than is the case for other firms, as regards both their credit lines and their bank deposits. Such a benefit might also occur if firms' portfolio of financial assets contains a given (high) proportion of assets whose issuers have been contributing to greening the economic system. In the contrary case, the firm might have to pay a penalty rate of interest on its borrowing and might earn a lower rate of interest on its bank deposits.

Such an incentive-based interest-rate policy could also affect the so-called "asset-price channel" of the monetary policy transmission mechanism. Firms as well as households whose portfolio contains "brown" financial assets will observe a reduction of the prices for these assets, as a result of a lower demand for them induced by central bank's policy as explained above. The opposite will occur for "green" financial assets, whose prices will increase because of a higher demand for them, particularly as soon as the return on this kind of assets increases after the ecological transition. Those firms whose portfolio contains "green" financial assets will be thereby in a position to borrow at lower rates of interest, both from the banking sector and across financial markets, since lenders know for sure that these firms are supported by the general government sector and the central bank, and moreover have high-yield financial assets in their portfolio to guarantee that they will be able to repay their debt at maturity.

Households could also benefit from such an "asset-price channel" in so far as they have invested in an environment-friendly real-estate object, which central banks may support through an interest-rate policy as described above. In such a case, indeed, the price of an environment-friendly flat or house can be higher, inducing the homeowner to consume more as a result of the so-called "wealth effect". Consumption will be supported thereby in an ecologically sustainable way that is likely to increase the employment level, too.

An analogous effect could result from the so-called "exchange-rate channel": if policy rates of interest are reduced for environment-friendly economic activities, and if these activities are prominent in the country's gross domestic product, the exchange rates of the national currency could depreciate, thereby inducing a higher volume of exports in an expansionary mood for the country's economic system. Of course, such an increase in exports should go along with the development of environment-friendly transports of those goods that are sold abroad, thus reducing global warming across the world.

4 Conclusion

Monetary policy strategies can support a green new deal for an ecological transition that has become dramatically urgent. Central banks can contribute to make the world a better place to live in, as regards climate conditions as well as financial stability. They are able to induce both firms and consumers to behave in an environment-friendly way in regard to economic decisions that affect the agents' well-being. This requires setting the policy rates of interest with regard to the positive or negative impact of economic activities on the environment. Such a monetary policy strategy will also have an impact on financial markets: asset prices will move in line with the ecological sustainability of the underlying firms, while financial institutions will have an interest in complying with some environment-friendly criteria for sustainable finance.

The favourable impact of monetary policy strategies on the environment will be much enhanced if the public sector issues green government bonds to finance, through public debt, an ecological transition that aims also at full employment. In this perspective, the general government sector should apply the pay-as-you-use principle, to wit, taxpayers should contribute to finance public debt as far as each generation of them benefits from public investments financed through the issuance of government bonds. If so, monetary policy strategies will have a much bigger impact on greening both economic growth and the whole economic system than would occur if the public sector is required to have a balanced budget as regards its capital account.

All in all, economic and financial systems will both contribute to and benefit from higher social cohesion if monetary policy strategies support a green new deal. The world will then become a better place to live in, owing to the implementation of a sustainable economy over the long run.

References

Bolton, P., M. Després, L.A. Pereira da Silva, F. Samama and R. Svartzman (2020), *The Green Swan: Central Banking and Financial Stability in the Age of Climate Change*, Basel: Bank for International Settlements, available at www.bis.org/publ/othp31.pdf (last accessed on 31 May 2020).

Campiglio, E. (2016), "Beyond carbon pricing: the role of banking and monetary policy in financing the transition to a low-carbon economy", *Ecological Economics*, **121**, 220–30.

Campiglio, E., Y. Dafermos, P. Monnin, J. Ryan-Collins, G. Schotten and M. Tanaka (2018), "Climate change challenges for central banks and financial regulators", *Nature Climate Change*, **8** (6), 462–68.

Carney, M. (2015), "Breaking the tragedy of the horizon—Climate change and financial stability", Speech at Lloyd's of London, London, 29 September, available at www.bankofengland.co.uk/-/media/boe/files/speech/2015/breaking-the-tragedy-of-the-horizon-climate-change-and-financial-stability.pdf?la=en&hash=7C67E785651862457D99511147C7424FF5EA0C1A (last accessed on 31 May 2020).

Dafermos, Y., M. Nikolaidi and G. Galanis (2018), "Climate change, financial stability and monetary policy", *Ecological Economics*, **152**, 219–34.

Dafflon, B. (1996), "The requirement of a balanced local budget: theory and evidence from the Swiss experience", in G. Pola, G. France and R. Levaggi (eds), *Developments in Local Government Finance: Theory and Policy*, Cheltenham, UK and Brookfield, WI, USA: Edward Elgar, pp. 228–50.

Danmarks Nationalbank (2019), "Climate change can have a spillover effect on financial stability", *Danmarks Nationalbank Analysis*, No. 26.

D'Orazio, P. and L. Popoyan (2019), "Fostering green investments and tackling climate-related financial risks: which role for macroprudential policies?", *Ecological Economics*, **160**, 25–37.

Eisner, R. (1984), "Which budget deficit? Some issues of measurement and their implications", *American Economic Review*, **74** (2), 138–43.

Eisner, R. (1992), "Deficits: which, how much, and so what?", *American Economic Review*, **82** (2), 295–8.

Karagiannis, N. and J.E. King (eds) (2019), *A Modern Guide to State Intervention*, Cheltenham, UK and Northampton, MA, USA: Edward Elgar.

King, J.E. (2023), "Functional finance", in L.-P. Rochon and S. Rossi (eds), *The Encyclopedia of Post-Keynesian Economics*, Cheltenham, UK and Northampton, MA, USA: Edward Elgar.

Krogstrup, S. and W. Oman (2019), "Macroeconomic and financial policies for climate change mitigation: a review of the literature", *International Monetary Fund Working Paper*, No. 19/185.

Lerner, A.P. (1944), *The Economics of Control: Principles of Welfare Economics*, New York: Macmillan.

Rossi, S. (2015), "Structural reforms in payment systems to avoid another systemic crisis", *Review of Keynesian Economics*, **3** (2), 213–25.

Sawyer, M. (2016), "The role of fiscal policy", in L.-P. Rochon and S. Rossi (eds), *An Introduction to Macroeconomics: A Heterodox Approach to Economic Analysis*, Cheltenham, UK and Northampton, MA, USA: Edward Elgar, pp. 193–210.

Svartzman, R., D. Dron and E. Espagne (2019), "From ecological macroeconomics to a theory of endogenous money for a finite planet", *Ecological Economics*, **162**, 108–20.

Volz, U. (2017), "On the role of central banks in enhancing green finance", *Inquiry Working Paper*, No. 17/01.

"Economic Possibilities for Our Grandchildren" Revisited

Romar Correa

1 Introduction

The possibility that a future for our grandchildren might not exist did not cross the mind of Keynes. Indeed, with his brisk, no-nonsense approach to problems and his workmanlike way of offering solutions he would have rolled up his sleeves and contributed with cleverness to the inter-governmental task of constructing Plans to mitigate the worst of Climate Change. The other subject of our paper, the threat of secular unemployment posed by cloud computing and the like displacing labour, Keynes foresaw with a prescience common to all great thinkers. His obiter dicta here, however, were less sure than his derivation of involuntary unemployment and the policy stance of public works. The likely reason is that he could no more than sense the nature of technical change to come and the displacement of labour brought about by technical progress required the sort of dynamics he was not familiar with. His remarks in the paper under review resonate today. "The increase of technological efficiency has

R. Correa (✉)
Independent Scholar, Mumbai, India
e-mail: romarcorrea10@gmail.com

N. Karagiannis and J. E. King (eds.), *Visions and Strategies for a Sustainable Economy*, Global Institute for Sustainable Prosperity,
https://doi.org/10.1007/978-3-031-06493-7_6

been taking place faster than we can deal with the problem of unemployment". Again, we are "... afflicted with a new disease ... technological unemployment".

As redolent today as then is the mood of foreboding. "We are suffering just now from a bad attack of economic pessimism", Keynes reports (Keynes, 2010). The seniors in the establishment of the day are not mincing words. The characterisation of the state of affairs is not one of economic depression, however deep, because the pit is the beginning of an upturn. Instead, all descriptions of the state of the world refer to subpar activity as enduring. An early and constructive general equilibrium model was the "structural slump" of Edmund S Phelps. Recently, Lawrence Summers has revived Alvin Hansen's expression "secular stagnation" to describe the USA, at least. Professor Hansen's term embraced the implosion of productivity, innovation, the American spirit no less in the America of his time in the last century.

We offer a constructive response to the two aforementioned forecasts. With regard to the first, our assumption is that market solutions like the exchange of pollution certificates or confidence in the streaming of market-driven innovations in the future or debates around the social discount rate are way off the mark. The task before the state is to grab economies with both hands and impose the discipline of national income accounting principles. Coming to the latter concern, we can borrow the labour-leisure tradeoff of mainstream practice to assume that a life of leisure is preferable to work. The classicals mulled over a world where drudgery was abolished and the erstwhile working class was occupied in refined play. Jamie Morgan (2019) also refers to Keynes' grandchildren essay in arguing that the relations of production under capitalism have become a fetter on the embracing of the new technology. Automation should be welcomed in a context that clearly signifies liberation from work. Old modes of regulation with public, cooperative, commons and peer-to-peer metrics will have to be rediscovered. Jan Fagerberg and Bart Verspragen (2020) share our prognosis that climate change and the "fourth industrial revolution" will determine the shape of capitalism in future.

We will continue to use the writings of Keynes along with classical insights as a sounding board. In the next section, we reinstate the long run.

2 Abolishing Short Termism

Large swathes of the world are agricultural with small and marginal landowners engaged in tilling the land with primitive technology. Other countries have leapfrogged the manufacturing stage into services. The outcome is piecemeal and precarious work for a few, constrained by the maximands of the principal internet giants abroad. Manufacturing offers hope for an increase in output and employment and the dignity and respect that the proletariat would be able to command through organisation.

Using the sharp characterisation of Luiz Carlo Bresser-Pereira (2018) the classic triad of profits, capital accumulation and innovation has collapsed into a univalent logic of rents. The ruling class is a combination of the bourgeoisie and what he calls the technobureaucracy. Originally, the capitalist class comprised variously the merchants or industrialists or bankers. Today they subdivide into business entrepreneurs or rentier capitalists. The technobureaucrats are professionals and the upper-middle class in the private and the public sector. The critical input into the production process is organisational knowledge. It is the information required to develop science and technology as well as manage corporations and the state.

Other forces have been at play. Competition is not healthy and socially useful anymore but is ferocious with races to the bottom. The guilty in the 2008 financial debacle pleaded innocence as they were compelled by the forces of cutthroat competition to mix and match securities illegitimately. In short, the future has collapsed in the face of the exigencies of the present. With technology and the fungibility of finance, the arbitrage principle rules. Monies will move, as the overworked cliché goes, at the click of a button.

The classicals were steeped in historical dynamics and posited the essence of capital to accumulate over time. The pulls and pressures of the financial circuit as an alternative to investment in factories and labour was not fully developed. Some corrective steps are being taken in the agenda to excavate and learn from money in the production and circulation of commodities in the political economy of Karl Marx (Gander 2019). The realisation of value depends upon the investment plans of firms as well as the consumption plans of households. The monetary theory of value comprises the demand for capital goods or investment plans of the capitalist class as well as the supply of labour of the working class. The incentive to accumulate the former is of a level of abstraction different from the inherent nature of surplus value to expand. In that regard, a qualitative solution to the "transformation problem" does not exist. A "transformation" from values to prices cannot exclude money. There is no value without exchange and money is the concrete manifestation of the appearance of value. The alleged focus on the production process in Volume I of Capital gave way to the unity of production and circulation in Volume III. Now, commodities are no longer associated with surplus product but with average profit. Keynesians have become sensitive to the insensitivity of private investment in plant and machinery to standard cues only recently. A regime of permanent unutilised capacity cries out for theory. Arne Heise (2019) concludes that the investment function used in neo-Kaleckian and neo-Marxian models is wanting. It is in the nature of capital to accumulate but capitalists make their choices, on the margin, between investing, consuming and staying liquid. They mull over the prospects of profits and an independent opportunity cost given by a rate of interest. Worse, history has provided cases of the surpluses of the earliest class societies being dissipated in subaltern consumption (Laibman 2020). Today, a low-level equilibrium trap resides at the tension between ecological crisis and technological breakthrough.

We underscore the centrality of the arbitrage principle. Investment will flow in response to risk-weighted asset returns. For instance, in a massive study Jordà et al. (2019) show that the total returns on risky assets like residential real estate and equities have been similar with high real total gains. With safe returns like bonds and bills, on the other hand, the returns have been volatile and negative. The long-run rate has been negative so as to ease pressure on government finances. Secondly, the term liquidity is the most salient in the aftermath of crises. Salvage is depicted in a scramble for liquidity. However, the antithesis of liquidity is illiquidity, the commitment of funding to plant and machinery and the employment of workers. Keynes was clear in Chapter 12 of the *General Theory*. "Of all the maxims of orthodox finance none, surely, is more anti-social than the fetish of liquidity". ... "It forgets that there is no such thing as liquidity of investment for the community as a whole". Again, "There is no clear evidence from experience that the investment policy which is socially advantageous coincides with that which is most profitable" (Keynes, 2013b).

We summarise this section with the appraisal of James Crotty (2019) which also sets the stage for our next section. Over the long haul, a capitalist economy will stagnate with occasional episodes of growth. Investment of the required scale will not be forthcoming. The government will have to take charge, listing out and prioritising projects which are both profitable and socially desirable. The composition of capital goods is important in closing the gap between aggregate demand and aggregate supply at the point of full employment.

3 Reinventing the State

A sophisticated engagement of departments of the state with the private sector would be founded on control and conflict. The ham-handed ceilings and floors on prices and quantities of yesteryear have been unlearned and these entities are now strategic variables. Secondly, the entry of the state does not mean nationalisation. In fact, nationalisation might not be antithetical to fascism, variants of which are a blight on nation-states all over the world. For instance, Mussolini's New Towns (Città di Fondazione) in Italy was a major infrastructure investment which was an important component in the fascist propaganda machine (Carillo 2019). The endeavour was successful in drawing electoral support for the party. Public spending as a plank on which to support non-democratic regimes was inspired by the cities of the Roman empire. The strategy was to galvanise the people on the lines of the collective consciousness of the ancient Romans. Vestiges remain in the longing for a strong leader, for nationalism, for racism, the central tenets of Fascism. Hitler believed in a nationalised health service and two copies of William Beveridge's famous report of 1942 which ushered in the post-war welfare state, richly annotated, were found in his bunker (Standing 2020). In this context, we report an alternative historical path that might have been trodden if a "proper" Keynes response to a deep recession had been adopted in the Lautenbach Plan of 1931 (Feld et al. 2021). The Plan, designed to address large-scale unemployment in an emergency manner, consisted of two planks, a public works program and nominal price and wage cuts. The former included the expansion of roads and railways. The finance, consisting mainly of a wage bill, would be provided by promissory notes issued by the national railway company, a form of implicit money financing. The latter consisted of the breaking of cartel prices by the government and the attendant flexibility of wages so as to restore

German competitiveness. The Plan was not accepted quickly enough to prevent political collapse and the victory of National Socialism. No more eloquent connection between nationalisation and fascism will be found than in the writing of Friedrich Engels (1970). Uncritical nationalisation is not socialism, he wrote. Bismarck assumed ownership of the main Prussian lines on behalf of the state only in order to have them at his command in the event of war and to indoctrinate the railway employees as "voting cattle" as well as create a source of monies independent of the constraint imposed by Parliament. Then, in a display of wit for which he and Marx were famous, "Certainly if the taking over by the State of the tobacco industry is socialisitic, then Napoleon and Matternich must be numbered among the founders of Socialism". Again, "… seriously proposed by a sly dog in Frederick William II's reign, the taking over by the State of the brothels".

In a remarkable leap forward centuries to our concerns in this paper, he envisaged capitalism evolving to conditions wherein the means of production develop to a level where the existing contradiction, the removal of their quality as capital, demands the entry of the state for the purpose of growth. "At first, the capitalistic mode of production forces out the workers. Now, it forces out the capitalists, …" Technical change operates "behind our backs" and even destructively as long as we do not seek to grapple with it. Once we master it, we can bend it to our will. Thereby, the anarchy of capitalist production gives way to the social regulation of production according to a Plan, serving the needs of society and of individuals. The components are on the one hand "direct social appropriation" for the maintenance and development of the means of production and, on the other, "direct individual appropriation" for the provision of basics to the working class.

We rely, once again, on Keynes for a foretaste of what are now called public–private partnerships. In Chapter 24 of the *General Theory* we find … "comprehensive socialisation of investment … need not exclude all manner of … devices by means of which public authority will co-operate with private initiative" (Keynes, 2013c). The state must determine the time profile of output and rates of return according to a Plan. "It is not the ownership of the instruments of production which it is important for the State to assume. If the State is able to determine the aggregate amount of resources devoted to augmenting the instruments and the basic rate of reward to those who own them, it will have accomplished all that is necessary". The monetary and fiscal authorities differ from representative agents at least with regard to time horizons. The former does not have continuously to mark to market. They do not have to maximise payoffs. They are not symmetric to private sector players and can enforce repeated play. The private sector will have a menu of options. Quick fixes must not dominate low and steady returns over a longer horizon. Otherwise put, the spectrum of financial avenues according to risk-return profiles can be constructed with private–public projects as points on a line. Reverting to Chapter 12 of the *General Theory*, "... the conclusion that to make the purchase of an investment permanent and indissoluble, like marriage, except by reason of death or other grave cause, might be a useful remedy for our contemporary evils". Finally, "In the case of very long-term investments, the risk can be ... shared between (the investor and the owner) ... the risk being outweighed in the mind of the owner by the advantages of continuity and security of tenure". In the end it is, ... "the State, which is in a position to calculate the marginal efficiency of capital goods on long views and on the basis of the general social advantage, taking an ever greater responsibility for directly organizing investment". ...

4 A Real Bonds Doctrine

Before we proceed to the novelty of our proposal, it is worthwhile to recall the non-novel connections between money and the state. Both heterodox and orthodox are joined in the proposition that a treasury that issues fiat currency debt and a central bank that issues fiat money in combination are powerful lenders of first resort (Akram 2020). A central bank is run-free. It cannot default on its liabilities. A government bond is a promise to pay government liabilities that are non-interest bearing at different points in the future. The central bank can influence the long-term interest rate on government bonds through the purchase of long-term government bonds and time- and information-contingent forward guidance.

The Real Bills Doctrine, recall, was a bill of credit contracted for by a commercial bank/central bank and a business entity for the purpose of meeting variable costs. The bill was of finite duration and was extinguished upon completion of the physical tasks. The difference in our scheme is that the principal and agent are the two complementary arms of the government, the central bank and the fisc. The two parts of the proposal, government investment alone, and people without work, are connected by a long-lived government instrument that is held by people and that provides them returns. We construct a government bond by fusing two ideas. On the one hand, an infinitely lived government bond is a perpetuity or consol promising the holder a dollar or euro, say, every year. We do not need to go far to seek elucidation. In the language of Chapter 11 of the *General Theory*, and substituting "man" with government, "When a government buys an investment or capital-asset, it purchases the right to a series of ... annuities, $Q_{1,}Q_2, \ldots, Q_n \ldots$" (Keynes, 2013a). The "prospective yield of that investment" over the series of Qs is the supply price of that asset, p_K, the price that, on the margin, would just induce a public sector enterprise to newly produce an additional machine or the "replacement cost". The marginal efficiency of capital (mec) is the discount rate which would equalise the two numbers. It is an equilibrium relationship. We denote the demand for capital goods by $I_{D,}$ prospective returns net of costs. On the other side is the yield on the existing capital stock, I_S. Since all calculations are being made by the government in our case of public investment, we assume that the central bank sets the discount rate. Lender's risk does not exist if the borrower and the lender are the same person. Double counting of the pure rate of interest must be avoided to give the minimum prospective yield which will induce the desired investment. Both the borrower and lender will consider the margin between the expected yield on the project and the borrowing cost. On the other hand, the present value of a bond is a definition. Once more with Keynes, "The rate of investment will be pushed to the point of the investment- demand schedule where the marginal efficiency of capital in general equals the market rate of interest. Q_r and d_r is the present value of £1 deferred r years at the current rate of interest. $Q_r d_r$ is the demand price of investment". In sum, $p_K I_S = (Q_r d_r) I_D = I_G$.

A historical fillip to the congruence of money and bonds is provided by the meticulous calculations of the historian Farley Grubb and his associates who have measured the transmutation of the real-backed government bills of British North America colonies into money (Wray 2019). Paper money was printed and disbursed as the wages of soldiers.

The market exchange value of the bill was decomposed as follows

$$MEV = (APV - RD)_t + TP_t$$

where APV is pure time-discounting, RD is default risk, TP is a transaction premium or pure "moneyness".

If $MEV \approx APV$, we have a low-risk asset.

If $MEV \approx APV - RD$, we have a risky asset.

If $MEV \approx TP$, we have a pure fiat currency.

The American colonies financed their activities by issuing their own monies of account like the Virginia pound or the North Carolina pound. The governments "spent the money into existence" and then collected the notes in taxes and finally burnt the tax reserves. No banks issued notes at that time and, therefore, taxes were called "redemption taxes".

The world of no work is a long way off and urgent attention is being paid to dignified work at a decent wage. Robert Rowthorn (2019) endorses the case for a Job Guarantee Programme unconnected with Universal Basic Income. All of a piece with our case made here, he supports government bonds to underwrite a Green New Deal with good-quality education and healthcare for all coupled with a massive programme of developing renewable energy. Finally, he underscores Keynes' advocacy of direct intervention by the government to lower the long-term rate of interest through the purchase of long-term government bonds. In the future, it is expected that automation will liberate people from jobs (Standing 2020). Time would be spent on work, not labour, as in care work and ecological work, on life-preserving labour rather than resource-denuding labour.

Climate change, on the other hand, is already ravaging the world. An ambitious goal to cut GHG emissions in the European Union by 65% in 2030 would require approximately 4.5% of GDP or 600 billion euros per annum (Storm 2020). Around 28 billion per annum of this funding is predicated on leveraging private sector capital by removing the risk for these investors whose investments are guaranteed in the EU budget. The severe danger here is that the finance is sought from interests deep in fossil-fuel investments. Instead of criminal liability, the Green New Deal offers a painless, profitable path into de-risked "green assets". These liquid assets will be in high demand. It is straightforward to expect that these riskless securities will be used to collateralise the cash pools of the global financial system. The destination is familiar. Speculation in unproductive and short-term exotic financial products will bloom. Besides, the share of profits in GDP in Europe has grown. The "degree of monopoly" has been increasing. The growing power of capital has come with a decreasing average profit tax rate. Raising the rate of corporate tax will not be unwelcome since corporations will benefit immensely from Green Investments.

The Plan is trammelled by the absence of working out the details of benefits to participants in the form of job creation and income distribution. Fiscal policy rules will continue to be a straitjacket. Secondly, a plus in favour of a Green New Deal is the creation of jobs in renewable energy generation, housing stock renovation and public transportation. Against this must be balanced the minus of the transition to a carbon-kind economy that will result in huge disinvestments in fossil-based industries, the evaporation of jobs in "brown" industries like coal mining and in fossil-fuel energy-producing and manufacturing industries like automobiles. The positive technological point here is that most carbon-intensive industries employ relatively few workers in proportion to their output. According to the ILO, the net effect is positive in terms of job creation for the EU during 2020–2030. No mention of smart zero-emission public transport. One million euros spent on public transportation creates 17–31 jobs.

The business-as-usual model would be disastrous for implementing a climate finance plan because the quantum of redirected investment for giving the world a fighting chance of limiting to 2 °C global warming is of the order of \$0.73 to \$1.3 trillion annually till 2035 with two-thirds being invested in developing countries (Aglietta and Coudert 2019). The Anthropocene, thus, is better described as the Capitalocene which is an irreversible transformation of the planet for the worse. Coal and oil will have to be phased out in energy production. Renewable resources are multiple and local. Climate change was defined at the Paris conference to be a global but differentiated public good. The measure of value would have to be the "social price of mitigation action" which is the tutelary price called upon to undertake huge, highly risky investments with long gestation lags and massive upfront costs. This notional price used in computing present-value returns would be the antidote to the negative verdict on these projects pronounced by the discount rate of market calculations. Long-term investors like public investment banks and "responsible" institutional investors are the need of the hour insofar as their asset portfolios match the long-term maturity of their liabilities.

Our concerns are expressed in the language of Stock-Flow-Consistent (SFC) models below.

5 An SFC Representation

Yet another reference to the grandchildren paper appears in the context of a comparison with the political economy of François Perroux (Fèvre 2019). Mainstream economics was found wanting in not considering abstract spaces which we interpret as general accounting matrices. Perroux sought purely monetary expressions for price and cost. In his view, Marshall and Keynes worked against the classical assumptions that investment choice only concerned private investment, that it was costly, and that it was the outcome of an individual decision.

We compare and contrast two economy-wide balance sheets following Godley and Lavoie (2007). The first is driven by "real" activity, notably by public sector firms employing capital to produce output. Total output, Y, is divided into consumption goods, C, and investment goods, I_G. In short, $Y=C+I_G$. We do not include the other standard categories like government consumption, G, and taxation, T, only to sharply differentiate our product from the familiar matrix. Unsubscripted entities are current variables, those with a -1 subscript pertain to "last year". Private investment is absent because it is absent in national income statistics all over the world! Indeed, the private sector represented below can be regarded as close to the wage goods sector of the classicals with basic consumption and investment in that Department, and returns on participation in government enterprise substituting for the wage bill. Government bonds are denoted by B and the rate of returns on bonds by r_b. Observe that in the absence of wages, national income decomposes into profits: $Y=F$.

A "real" economy

	Households	*Public Firms*		*Central Bank*	*Government*	$\sum$
		Current	*Capital*			
Consumption	$-C$	$+C$				–
I_G						
Investment		I_G	$-I_G$			0
GDP (memo)		Y				
Profits		$-F$	$+F$			0
Income	$+r_b L_{-1} B_{-1}$		$r_b L_{-1} B_{-1}$	$+r_b L_{-1} B_{-1}$	$-r_b L_{-1} B_{-1}$	0
Change in bonds	$-r_b L \Delta B$		$+r_b L \Delta B$	$(-\Delta M)$	$+r_b L \Delta B$	0
$\sum$	0	0	0	0	0	

We see a rare and resplendent exhibition of quadruple-entry book-keeping which is the defining trait of SFC economics. All characters are on the stage in all acts. In the present, households consume by availing of their share of public wealth or equity holdings in public sector firms by availing of their interest income earned. The accounts are held with the central bank. Independent of this discussion, with Central Bank Digital Currency (CBDC) just around the corner, this expectation is rational. All income is not consumed. Households would be willing takers of a fresh infusion of government bonds in the current period. The notation ΔM instead of the flow of the tranche of bonds is used as a catchall of the elision between MEV, APV, RD and TP introduced earlier. The box is bare. In a minimal extension, nationalised (not in a traditional fashion) banks would be adjuncts to the central bank. The matrix is a static representation but the photographs taken at points of time can be fast forwarded. The matrix dynamics is stable because none of the elements are arbitrary or unbounded but are grounded by the physical output returnable by the public sector enterprises.

The chart can be modified in familiar terms to display "quantitative easing for the people" (Stockhammer 2020). In a recession that will not go away unemployment hysteresis implies not only an increase in unemployment but also a hike in the NAIRU. The system is path-dependent. The economy has a memory. In that case, fiscal multipliers are strictly positive and government spending has long-lived effects. An increase in demand creates an increase in supply. Central banks can finance government expenditure directly with finance through an intermediary like a National Investment Bank.

For Keynesians in particular, the task is to direct investment in productive directions (Palma 2020). Lukewarm growth is as much about the composition of effective demand as it is about its level. Increasing inequality pushes demand away from its most productive uses. The path ahead on Main Street is still dark and the WHO has warned of the economic epidemic becoming a pandemic. Yet, expectations of further monetary and fiscal stimulus are entrenched and so all bets on an unabated rise in share prices are on. The derivative books of most US banks are larger than the time when Bear Stearns had to be rescued. At the end of 2017, five of the biggest US banks had derivatives books worth US$157 tn, twice global GDP.

The contours of an alternative, stylised "financial" economy are not hard to shape, therefore. Activity and employment have been listless all over the world for over a decade. Investment is moribund. The working class has been broken even earlier. The prospect of robots and AI substituting for work is real and has been modelled by dropping the wage bill from our accounts. At the same time, the so-called middle class swells and is not opposed to the milieu. In the absence of a buoyant job market, that can only mean that it receives handsome returns from financial markets. Banks in the traditional sense are near extinction as attractive business projects are hard to find and as households find more attractive avenues in which to park their savings. Concurrently, different forms of financial entities appear (and disappear) luring firms as well with the promise of smart, short-term gains. Non-financial firms become financial firms. Their offerings are e_b at the price p_{eb}. These can signify plain vanilla instruments or more esoteric securities and are held by households, e_{bh}, and firms e_{bf}. For the sake of brevity, we drop I and the share-financing by firms of smart machines in what follows. Our aggregate identity now is $Y=C$, C here stands for the consumption of rentiers. Profits are off-balance sheet.

A financialised economy

	Rentiers	Financial firms: Current	Financial firms: Financial Institutions	$\sum$
Consumption	$-C$	$+C$		0
GDP (memo)		Y		
Revaluations	$+\Delta p_e b.e_b h$		$-\Delta p_e b.e_b$	0
Changes in equities	$-\Delta e_b h.p_e b$		$+\Delta e_b.p_e b$	0
$\sum$	0		0	

The driver of the financialised economy can be seen to be the revaluation term or the capital gains item in the balance sheet. The doom loop only works on the assumption of the continuous upward movement of the price of financial assets. Reality is a check in feverish innovations on Wall Street proceeding unabated in the face of the caving in of Main Street. The plus and negative mirror items are hoist by their own petard of optimism by rentiers and financial firms. The music will not play forever but until it stops there is much dancing.

6 Discussion

Returning in the end to the foreboding of the beginning, not a few see countries described by the economics of trade and consumption alone (Therborn 2020). Producers, workers, social relations are not to be found. Activity consists of the middle-class employment of teachers, health personnel and public service employees. Middle-class environmentalism is in conflict with the rapacity of real-estate developers, extraction corporations and pollution generators.

Joel Mokyr has documented how the foundations of modern economic growth, laid in the period 1500–1700, were driven by a transformation in beliefs, values, preferences, summarised in the word "culture" (Gehrke 2018; Spolaore, 2019). These preceded the Industrial Revolution and encouraged the development of "useful knowledge" from an engagement with the natural world. A fruitful combination of *episteme* and *techne* was ignited. The former consists of a society's understanding of natural laws based upon observation and testing. The latter consists of the available technological blueprints. The deep and virtuous connection between the two was fostered by the Baconian ideal that turned natural philosophy to the service of commerce and manufacturing. Ideology must not be confused with institutions. Supposedly "optimal" institutions might not deliver. Institutions do not seem to have played any role in the Industrial revolution. Thus, governments have enforced property rights for aeons without any visible positive outcomes. Dennis Snower (2019) has reminded us that growth over the past three hundred years was generated primarily by cooperation. The need of the hour is the transition from a "welfare state" to an "empowering state". Such a state would educate and train people in spheres where humans cannot be substituted by machines. These are skills involving sociability, noncodifiable creativity, and the ability to take on peculiarly human responsibilities. Abilities like sympathy, compassion and conceptualisation, can be taught and tested.

7 Conclusion

Our grandchildren will surely inherit an earth where the production of goods and most services like education and medical care will be delivered by machines. Our conjecture is that the scale and technological demands and costs of the new mode of production and exchange and circulation cannot be addressed by markets alone. Governments will have to ensure domains where their interests are aligned with the people and pursing self-interest is best served by policy coordination. Predatory finance, for instance, knows no national boundaries and walls will have to be erected to defend the consumption and care of citizens from the fatal beguile of its siren song. People will be in possession of society participation certificates. They would own shares in public enterprises that are underwritten by the government. Government bonds would circulate as fiat money which would be completely backed by ecologically kind goods and services. Elements like capital saturation and so on are vestiges of manufacturing technology and diminishing returns might not apply to cloud computing. We have demonstrated the discipline of double-entry bookkeeping applying to the snapshot of our economy as it spools over time. Transversality or no-Ponzi-games conditions do not apply.

References

Aglietta, Michel and Virginia Coudert, 2019, The Dollar and the Transition to Sustainable Development: From Key Currency to Multilateralism, CEDII November 26, Policy Brief May 2019

Akram, Tanweer, 2020, A Note Concerning Government Bond Yields, Levy Economics Institute of Bard College Working Paper No 977
Bresser-Pereira, Luiz Carlos, 2018, Rentier-financier capitalism, Sao Paolo School of Economics Working Paper 477
Carillo, Mario Francesco, 2019, Fascistville: Mussolini's New Towns and the Persistence of Neo-Fascism, MPRA Paper No 96236, posted on October 8, 2019
Crotty, James, 2019, *Keynes Against Capitalism*, London & New York: Routledge
Engels, Friedrich, 1970, *Socialism: Utopian and Scientific*, Moscow: Progress Publishers
Fagerberg, Jan and Bart Verspragen, 2020, Innovation-Diffusion, the Economy and Contemporary Challenges: A Comment, *Industrial and Corporate Change*, 29, 4, 1067–1073
Feld, Lars P., Ekkehard A. Köhler and Daniel Nientiedt, 2021, The German Anti-Keynes? On Walter Eucken's Macroeconomics, *Journal of the History of Economic Thought*, 48, 4, 548–563
Fèvre, Raphaël, 2019, Georges Bataille, François Perroux and French Critiques of the Marshall Plan, OSF Preprints, July 6, 2019
Gander, Sascha, 2019, The European Economic Crisis from 2017 Onwards in the Context of the Global Crisis of Overproduction of Capital-A Marxian Monetary Theory of Value Interpretation, Institute of International Political Economy, Berlin, Working Paper No 125/2019
Gehrke, Christian, 2018, Joel Mokyr's *A Culture of Growth*: A Book Roundtable, *The European Journal of the History of Economic Thought*, 25, 6, 1493–1536
Godley, Wynne and Marc Lavoie, 2007, *Monetary Economics*, Palgrave Macmillan.
Heise, Arne, 2019, Neo-Kaleckian and Neo-Marxian Regime Research: A Promising Scientific Research Programme or a Scientific cul-de-sac?, Zentrum für Ökonomische and Soziologische Studien, Discussion Paper No 74
Jordà Òscar, Katharina Knoll, Dmitry Kuvshinov, Moritz Schularick and Alan M Taylor, 2019, The Rate of Return on Everything, 1870–2015, *The Quarterly Journal of Economics*, 134, 3, 1225–1298
Keynes, John Maynard, 2010, "Economic Possibilities for Our Grandchildren" in Essays in Persuasion, Palgrave Macmillan, 321–332
———, 2013a, "The Marginal Efficiency of Capital", in *The Collected Writings of John Maynard Keynes Volume VII The General Theory of Employment, Interest and Money*, New Delhi: Cambridge University Press
———, 2013b, "Chapter 12: The State of Long-Term Expectation", in *The Collected Writings of John Maynard Keynes Volume VII The General Theory of Employment, Interest and Money*, New Delhi: Cambridge University Press

———, 2013c, "Chapter 24: Concluding Notes on the Social Philosophy Towards Which the General Theory Might Lead", in *The Collected Writings of John Maynard Keynes Volume VII The General Theory of Employment, Interest and Money*, New Delhi: Cambridge University Press

Laibman, David, 2020, China: In the Perspective of Historical Materialism, *Science and Society*, 84, 2, 171–203

Morgan, Jamie, 2019, Will We Work in Twenty-First Century Capitalism? A Critique of the Fourth Industrial Revolution, *Economy and Society*, 48, 3, 371–398

Palma, José Gabriel, 2020, Finance as Perpetual Orgy: How the 'New Alchemists' Twisted Kindelberger's Cycle of "Manias, Panics and Crashes" Into "Manias, Panics, and Renewed Manias", Faculty of Economics, University of Cambridge, Cambridge Working Paper in Economics 2094

Rowthorn, Robert, 2019, Keynesian Economics-Back from the Dead? The Godley-Tobin Lecture, Centre for Business Research, University of Cambridge

Snower, Dennis, 2019, A New Societal Contract, Economics: The Open-Access, *Open-Assessment E-Journal*, 13, 37, 1–13

Spolaore, Enrico, 2019, Commanding Nature by Obeying Her: A Review Essay on Joel Mokyr's *A Culture of Growth*, NBER Working Paper 2606

Standing, Guy, 2020, Interview with Mitja Stefancic, World Economic Association Commentaries 10, 2, May 2020

Stockhammer, Engelbert, 2020, Post-Keynesian Macroeconomic Foundations for Comparative Political Economy, Post-Keynesian Economics Society Working Paper 2022

Storm, Servaas, 2020, The EU's Green Deal: Bismarck's 'What Is Possible' Versus Thunberg's 'What Is Imperative', Institute for New Economic Thinking Working Paper No 117

Therborn, Göran, 2020, Dreams and Nightmares of the World's Middle Class, *New Left Review*, 124

Wray, L Randall, 2019, Alternative Paths to Modern Money Theory, *Real-World Economics Review*, 89, 5–32

Costing the Green New Deal in the United States: The Modern Money Theory Approach

Yeva Nersisyan and L. Randall Wray

1 Introduction

Advocates of the Green New Deal (GND) strive to change the way we approach a variety of problems facing society: climate change and destruction of our natural environment, rising inequality, and an economy that leaves too many with inadequate access to food, shelter, healthcare, and affordable education. They see these problems as linked, and so insist on tackling them with an array of programs that have hitherto been seen as disconnected: a carbon-neutral energy policy and reversing climate change; universal single-payer healthcare; student debt relief and free

This chapter is a shorter version of the Working Paper No. 931 that was published by the Levy Economics Institute of Bard College in May 2019.

Y. Nersisyan (✉)
Economics, Franklin and Marshall College, Lancaster, PA, USA
e-mail: ynersisy@fandm.edu

L. R. Wray
Levy Economics Institute of Bard College, Annandale-on-Hudson, NY, USA

N. Karagiannis and J. E. King (eds.), *Visions and Strategies for a Sustainable Economy*, Global Institute for Sustainable Prosperity,
https://doi.org/10.1007/978-3-031-06493-7_7

public college; prison reform; ending “forever wars”; increasing care for the young, sick, and old; and the job guarantee.

Many supporters and proponents of the GND are worried that large—perhaps confiscatory—tax hikes will be needed to “pay for” it. They typically warn of the high financial costs, and hence of prospective dangerously high government deficits. From the perspective of Modern Money Theory (MMT), these arguments are beside the point. A sovereign government’s finances are not like the budgeting by households and firms; the government uses the monetary system to mobilize the *nation’s* real resources and to move some of them to the pursuit of the public purpose. Whatever the financial costs, we already have a financial system that can handle them.

What is less certain is whether there are suitable real resources that can be mobilized for a GND. To determine that, we need to do an accounting of available resources (including potential excess capacity and resources that can be shifted away from existing production) and to weigh those against what will be needed. If we can identify technologically feasible projects that would achieve the GND’s goals and the resources to devote to these projects, then we can arrange for the financing of the programs. Only after a careful resource accounting can we address the question of whether taxes and other means might be needed to reduce private spending to avoid inflation as the GND is phased in.

This contribution summarizes findings presented in detail in our Levy Institute working paper (Nersisyan and Wray 2019). Here we use estimates of resource sources and uses that we generated from our literature survey and from our own calculations. Interested readers are referred to the working paper for more details on our methodology and calculations. We hope our attempt spurs further research to carefully assess resource availability to compare against the resource needs of full implementation of the GND we outline here.

Our approach closely follows that of John Maynard Keynes in *How to Pay for the War* (1940). In order to determine the net resource requirement for the GND, we evaluate the main GND projects to gauge whether they would be a source or a use of resources. We largely follow Keynes’s method, which is to mostly use monetary measures (dollars in our case) as proxies for resource quantities. Ideally, we would use a measure of real productive capacity—the ability of a resource to produce output—but as we have heterogenous resource inputs and heterogenous outputs, this is difficult. As such, the estimates can provide only a very rough guide. Uncertainties remain concerning the technologies, the quantity of resources needed, the suitability of the resources that can be made available, the political feasibility, and the capacity of our democracy to successfully face the challenges ahead. But we believe that our approach, informed by MMT and Keynes, provides more guidance about the question of the GND's "affordability" than the conventional approach of merely adding up the dollar "costs" of GND projects. Assessing economic feasibility of the GND needs to focus on technological know-how and resources, not on the dollar costs.

2 Affordability: The MMT Approach

MMT maintains that a sovereign government is not financially constrained; it spends through electronic entries in bank accounts and can neither run out of them nor save them for the future (Wray 2015; Nersisyan and Wray 2019). What should constrain the spending of a sovereign government is the nation's available real resources. Excessive spending, therefore, creates problems not in terms of higher government deficits and debt, but in terms of true inflation. Similarly, taxes are used not to finance government spending, but to withdraw demand from the economy, creating space for government spending to move resources to the public sector without causing inflation.

The more sensible critics of MMT accept the argument that we can financially afford the GND but insist we need to raise taxes more or less in line with spending to avoid causing inflation. They go on to assert that increasing spending without raising taxes is a particularly dangerous practice as it injects excess money into the economy that will cause inflation. For instance, Brad DeLong has argued, in reference to boosting federal government spending, that it is unreasonable to believe that "the US can have Swedish levels of government spending without Swedish levels of taxation."[1]

[1] See Wray (2019). DeLong is wrong. American workers already pay Northern European levels of "taxation" if you use an inclusive measure: all the mandatory deductions from American paychecks (pensions and health insurance, including the Obamacare mandates, called "non-tax compulsory payments," or NTCPs) add up to a greater burden than what our rich peer countries' workers pay. For comparison purposes, Canadian workers pay an effective "tax" rate (including NTCPs) of just 11.5 percent; in Denmark they pay 26.7 percent; in Norway 32.4 percent; in Sweden 38.3 percent; and in the United States a whopping 43.2 percent (Bruenig 2019). In sum, we already pay higher

The argument that taxes must be increased along with the GND presumes that, first, implementing the GND will lead to a large increase in the demand for resources, and, second, that raising taxes is the best way to relieve pressure on demand. We believe that those arguing for tax hikes have not carefully assessed the inflation potential of the GND, since there have been no studies of the GND that look at resource demands and savings. Instead, they rely on estimates that merely sum up the estimated *dollar* costs of GND programs, presuming this equates to additional government spending that requires taxes to "pay for" it. Whether or not we need tax hikes depends on the net increase of demand on the nation's resources and not on some scary prediction of trillions of dollars of government red ink. We provide a first attempt to quantify the net resource demands in order to obtain some idea of the reduction of current aggregate private demand (consumption and investment) that might be required to release resources to the GND efforts—that allows us to deal with what would otherwise create an inflationary excess demand.

It is not a foregone conclusion that the net increase will be beyond our available capacities. The GND will in many areas reorient resource use, cutting environmentally and socially destructive uses of resources and shifting them to better use. Moreover, we have substantial unused capacity, and we will, along the way, further increase capacity. That is obvious in Euroland, which is probably operating 25 percent or more below full capacity, but even the United States has substantial excess capacity, in spite of claims that we are at full employment. Over the past quarter-century, we have had to repeatedly lower our estimate for the natural rate of unemployment—the rate below which inflation is supposed to pick up—in every recovery, because inflation never arrives as unemployment falls. Most recently, in the immediate aftermath of the financial crisis, economists claimed that the natural rate of unemployment had increased to over 5 percent; yet the unemployment rate kept coming down, while inflation never materialized. Even the Federal Reserve has admitted that the inverse relationship between unemployment and inflation that underpins their policy decisions does not seem to exist anymore (if it ever did)

taxes than the Swedes; we just don't call them taxes, even though they are as mandatory as Swedish taxes. Take-home pay of Americans is already below that of Swedes, which is obvious to anyone who travels to northern Europe to envy the standard of living we do not enjoy.

(Li 2019). We will not know precisely where the true inflation barrier is until we ramp up aggregate demand.

Further, there is nothing "natural" about potential growth. We can raise our potential by prudent investments of our resources. Operating close to full capacity over a period will bring forth investment and more capacity—so potential output is to some degree determined by actual capacity use. Through the GND, we will also shift resources to better uses and (gradually) reduce the need to devote resources to dealing with the problems created by destructive processes (dirty production processes require us to devote resources to cleanup; high inequality forces us to devote resources to dealing with the consequences of poverty, idleness, inadequate access to preventive care and early treatment, social isolation, and crime).

To sum up, we need to obtain an inventory of the resources that can be made available to the GND projects to compare against the resources that will be required to implement the GND. This would include resources in excess supply plus those that can be released from uses that will be eliminated by adoption of the GND. This is the true cost of the GND, and it will allow us to get some idea of the magnitude of the reduction of aggregate demand necessary to avoid inflationary pressures. Admittedly, this is difficult, and we are only providing a first step, using gross measures of underutilized resources currently available, resources that can be shifted, and resources that will be needed by GND projects.[2]

To be clear, we do recognize that implementation of the GND will shift spending from the private sector to the government sector. However, unlike other studies that evaluate the GND, we do not assume that we need to raise taxes merely because government spending rises. Instead, our position is that taxes should be raised only if the additional spending creates inflationary pressures. In general, it makes no difference *in terms of inflation* whether the dollar spent to hire resources comes from the government or from the private sector. Thus, a shift in spending from the private sector to the government required by the GND need not be inflationary. However, if the GND does take us to and beyond full

[2] In this policy brief, we do not assess the technological feasibility of the part of the GND that is focused on reversing climate change. We will assume that the science and technology exist. We have no expertise in that area. Instead, we focus on resource availability.

employment of resources, we can adopt measures to counter inflation pressure, including raising taxes.

3 "Costing" the Green New Deal

The American Action Forum (AAF), led by the former Director of the Congressional Budget Office Douglas Holtz-Eakin, is the source of an oft-cited GND cost estimate of $93 trillion—a number various commentators and opponents of the GND have used to argue that it is wishful thinking at best and a terrible policy that will bankrupt our grandchildren at worst (Holtz-Eakin et al. 2019). However, this number is obtained by grossing up wild estimates of the individual costs of each component of the GND—without accounting for potential savings as we phase out wasteful and damaging uses of resources.

What we do instead is to consider the resource savings that will accrue from the various programs, as well as the resource costs of GND programs, to determine whether the GND would be affordable in terms of the *available real resources*. As we demonstrate, some parts of the GND are resource-creating while others are resource-saving, and it is highly misleading to simply tally the projected dollar costs. In what follows we briefly summarize our estimates for resource savings and needs for the main GND components.

3.1 *The Job Guarantee Path to Jobs for All: (and a Source of Resources for the GND)*

While there are many job guarantee (JG) proposals floating around, the Levy Economics Institute proposal is among the most ambitious (Wray et al. 2018). The Levy proposal includes paying $15 per hour plus generous benefits (at 20 percent of the wage bill, including Medicare-style healthcare and free childcare), plus an amount of spending equal to 25 percent of the wage bill to cover materials costs. Thus, the JG not only provides full employment but it also ensures an effective national minimum wage of $15 per hour—and this is accomplished whether or not $15 is the legal minimum.[3]

[3] The JG wage will become the effective minimum wage—assumed to be $15 per hour plus benefits. Without a JG, a legal $15 minimum wage may not be an effective minimum wage, because those who cannot obtain a job in the formal labor market will not receive

The Levy simulation of the JG puts the net annual impact on the federal government's budget at around $400 billion per year for the first 10 years; state budgets improve by $53 billion annually. The boost to GDP is around $560 billion annually, while the boost to employment is around 19 million new workers (15 million in the program, plus 4 million private sector jobs).[4] What is important here is the impact on resource use, not the budgetary impact on the federal government. The net $400 billion boost to federal government spending (about 2 percent of GDP) means that aggregate net wages have been increased by about that amount.[5]

While in financial terms the JG represents a cost, in real terms it is both a cost (it uses resources—mostly the increased consumption out of wages by employees over what they consumed before they got jobs) and a source of resources for GND projects. In particular, JG workers can be employed for many of the GND projects. By design, JG projects would be able to utilize labor with below-average skills and experience and the jobs would be labor-intensive, requiring little expensive capital investment or materials. These could include care services (care for the environment, community, and people) plus small construction and retrofitting projects (making homes more energy-efficient, for example). Clearly, these workers would not be used as skilled labor in major infrastructure projects, a core component of the GND.[6] Importantly, the JG would be directly employing labor resources that are not currently being used by the private sector.

that wage (they might remain unemployed or be forced to work in informal labor markets at less than the minimum).

[4] These are high estimates, since the Levy simulation does not include likely cost reductions, such as lower spending on social programs and the penal system that would result from poverty reduction through job creation. We assumed some budgetary savings from lower Medicaid spending and reduction of the Earned Income Tax Credit—as program workers would have higher incomes that would raise them above program thresholds.

[5] Also note that Medicare-style healthcare as well as childcare coverage is included in this simulation of a JG program. With Medicare for All, JG workers would get healthcare through that program. Hence the spending on the JG itself would actually be lower than the Levy estimates if the entire package of GND programs was adopted.

[6] The Levy report (Wray et al. 2018) discusses the importance of respecting prevailing wage legislation and avoiding competition with union labor. Further, most public infrastructure projects will continue to be undertaken through contracts with private firms—hence, would not be performed by the JG program.

Let us assume that half the JG workers are in projects devoted to "greening"; the other half are devoted to what Pavlina Tcherneva (2018) calls "care for community" and "care for people" projects: service projects related to senior and youth care, teacher's helpers, neighborhood and park cleanup, artistic projects, and so on. Thus, the JG program can provide resources needed for green projects in an amount equal to 1 percent of GDP (and resources equal to another 1 percent of GDP for other care services).

To conclude, we count the JG as using an amount of resources equal to 2 percent of GDP. However, we count it as supplying an amount of resources for green projects equal to 1 percent of GDP—hence, the net cost in terms of resource use is 1 percent of GDP.[7]

3.2 *Renewable Energy and Energy Efficiency*

For our estimates of resource use required to transition fully to renewable and efficient energy sources, we use estimates of "greening projects" (100 percent renewable energy, a smart power grid, and upgrading and retrofitting buildings) from a variety of sources, including a 2014 report from the Center for American Progress (CAP) and the Political Economy Research Institute (PERI) (Pollin et al. 2014); the AAF (Holtz-Eakin et al. 2019); a plan by Mark Jacobson from Stanford (Jacobson et al. 2017); one by Geoffrey Heal of the Columbia Business School (Heal 2017); and Senator Bernie Sanders' plan (Bernie 2020, n.d.). Table 1 summarizes the estimated costs of "greening" the economy, using an average of various proposals, including the one by the conservative American Action Forum, as well as Bernie Sanders' plan, which is the most expensive (and detailed). An average of all the estimates is $1104 billion annually, or 5.66 percent of 2017 GDP, which is what we use in our estimates going forward.

In transition to a zero-emissions system, however, we would stop all investment in the nonrenewable sector and gradually dismantle existing nonrenewable capacities. According to the Bureau of Economic Analysis,

[7] We have chosen not to directly count the contribution of JG workers in "care" services as a net resource because we do not include an estimate of the resource costs of the care services. Thus, we are assuming that the JG care services essentially "pay for themselves" in terms of resource use; however, the JG care workers will consume 1 percent of GDP, so they are treated as a GND resource cost.

Table 1 Summary of various "greening" proposals

	\$ Billion	*% of 2017 GDP*
Jacobson	1250	6.41
AAF	1030	5.3
CAP/PERI	1021	4.98
Heal	580	2.97
Sanders	1638	8.4
Average	**1104**	**5.66**
Fossil Fuel Investment	−127	−0.65
Net Cost	**977**	**5.01**

private investment in "mining exploration, shafts, and wells" was \$140.9 billion in 2018, or about 0.7 percent of 2017 GDP (BEA 2019). We estimate that about 90 percent of that, or \$127 billion (0.65 percent of GDP), is an investment in petroleum and natural gas.[8] To get the net cost of "greening," we would then need to subtract the sum of our current spending on nonrenewables from our estimates. Furthermore, since we are already investing in renewables, that cost may need to be subtracted as well. However, since it is not always clear whether the proposals we examined are estimating net or total costs of greening, we will err on the side of caution and will not subtract current renewable investment.

3.3 *Universal Single-Payer Healthcare*

The universal single-payer system, or Medicare for All (M4A), is deemed the most expensive GND program. However, as we explain below, M4A is actually a source of resources in real terms. The US healthcare system is notorious for its high costs and below par outcomes. We already spend 18 percent of GDP on healthcare, and that is projected to reach 20 percent soon. This is approximately twice as much as our peers—other rich, developed capitalist countries—with no discernably better health outcomes

[8] While Table 5.3.5 (BEA 2019) does not provide a further breakdown between petroleum and natural gas and other mining, Table 5.4.5, "Private Fixed Investment in Structures by Type," does. According to that table, oil and natural gas investment comprised about 93 percent of total investment in structures, while investment in other mining was less than 7 percent. We are using a 90/10 breakdown in our estimates.

(and even worse on a number of measures).[9] Our excessive spending when compared to that of our peers can be attributed to the use of for-profit insurance to pay for healthcare, higher pharmaceutical and provider costs, and higher administrative costs. Study after study has confirmed that prices and administrative costs in the United States are out of line with those in the rest of the developed world and especially compared to countries that have some type of a single-payer system. Americans do not get more healthcare in real terms, but they pay more for it (see Nersisyan and Wray 2019 for a brief survey of literature).

The adoption of a single-payer system (replacing for-profit private insurers) would significantly reduce the resources devoted to our unusual way of paying for healthcare. It would eliminate the private insurance sector's participation, reduce employers' costs of administering healthcare plans, reduce the costs incurred by doctors and hospitals due to billing insurers as well as pursuing patients for uncovered cost, lower the costs of appealing denials, and cut costs associated with patients avoiding early treatment of diseases (because of the actual or expected out-of-pocket costs) that become chronic and expensive maladies. If M4A could control prices and lower administrative costs, we could spend significantly less on healthcare than we do currently, while expanding coverage to everyone. All else equal, if we were able to reduce our spending on healthcare to the level of our peers, we would be creating deflationary pressures, not inflation.

We estimate that in the short term M4A could save about 3.7 percent of GDP while providing healthcare to the whole population (Nersisyan and Wray 2019).[10] Even if we lowered healthcare spending by 3.7 percent of GDP, we would still be spending more on healthcare than all of our peers. We believe our estimates are just the savings possible in the short term. In the long term, increased use of healthcare could reduce spending on chronic diseases. With universal access, cost controls, and elimination of a highly inefficient private insurance system, the single payer could shrink US spending on healthcare by much more, bringing us in line with other rich countries at about 10 percent of GDP. However, for the

[9] The OECD country with the next highest share of health spending is Switzerland, which spends a little over 12 percent of its GDP on healthcare. Canada, on the other hand, spends 10.4 percent of its GDP on healthcare (OECD 2019).

[10] Others have estimated savings from 1.58 percent of GDP (Pollin et al. 2018) to 4.5 percent of GDP (Baker 2019a).

purposes of this analysis we will use our estimate of savings equal to 3.7% of GDP.

Some will object that the savings largely accrue to the private sector, while the government will face additional costs. While it is true that the distribution of spending between private and public sectors would change, there is nothing about government spending that necessarily makes it more inflationary than private spending—all else equal. If private spending on healthcare costs falls by more than the increased spending by government, the movement to single payer will be deflationary, not inflationary. Only a net increase in demand for resources would be inflationary. And if M4A is implemented as a part of the GND, what will matter is the overall increase of demand resulting from all of the GND. If healthcare savings offset increased resource use in other GND projects, then the net impact of the GND will not be inflationary.

3.4 End the Forever Wars

Some advocates of the GND (including Senator Sanders) have proposed to end the "forever wars."[11] This will reduce environmental destruction and free up resources for the GND. It is hard to calculate the total amount of resources devoted to the forever wars, as so much of the defense budget is hidden in other programs. The reported spending on defense is $716 billion, of which reported war spending is about 10 percent (say $72 billion). However, much more is hidden, and this figure does not include indirect costs (e.g., treating veterans and their families). As military spending does not lead to production of output for civilian consumption, it is akin to transfer payments in terms of its inflationary impact. There is probably at least $1 trillion of "unproductive" military spending annually. In 2008, Joseph Stiglitz projected the cost of the Iraq War alone would run about $3 trillion; two years later, he argued that

[11] "Congress made history recently by passing a resolution that cuts off U.S. support for Saudi-led forces in the civil war in Yemen. This is the first time since Congress originally passed the War Powers Resolution in 1973 that we have used it to call on the president to withdraw from an undeclared war ... It is time for Congress to ask whether, nearly 18 years after 9/11, we really want to continue to be involved in these wars for another 18 or more. According to a recent study by the Costs of War Project at Brown University, the War on Terror will have cost American taxpayers almost $5 trillion through Fiscal Year 2019. When taking into account future health care obligations for veterans injured in post-9/11 wars, the bill comes closer to $6 trillion" (Sanders and Lee 2019).

could be an underestimate (Stiglitz and Bilmes 2010).[12] Senator Rand Paul claims the war on terror has cost $6 trillion since 2001, or about $330 billion per year (Shane 2019). Senator Sanders has put the cost at $5 trillion already, with another trillion to be spent on healthcare for veterans over the future.

While there are great uncertainties surrounding estimates of the resources that can be saved by ending the forever wars, we will use a conservative estimate that $210 billion of savings could be realized annually (perhaps two-thirds of the spending on the war on terror) in terms of wasted resources, or approximately 1 percent of GDP annually. We expect that the savings would be lower in the early years but would gradually rise—probably significantly above 1 percent of GDP—as the longer-term costs of caring for veterans and their families would decline over time.

3.5 Achieving Net Zero Carbon Emissions in Agriculture

Global agriculture accounts for about a quarter of all greenhouse gas emissions. Hence, some supporters of the GND have recognized that achieving carbon neutrality will require major reduction of emissions by agriculture—and in particular, reduction of greenhouse gases produced by livestock. According to the United Nations Food and Agriculture Organization, global livestock production accounts for nearly 15 percent of emissions, "with beef and milk cattle production comprising 41 and 20 percent of the sector's total, respectively. Greenhouse gases also stem from feed and animal transportation, manure storage and processing, and cutting down forests to expand farmland for crops and grazing, particularly in developing economies" (Dmitrieva 2019). Obviously, agriculture has other significant impacts on the environment in addition to carbon emissions and loss of forests. Factory-farmed beef, pork, and fowl is a significant source of pollution of air and water resources. Finally, eating large quantities of meat is not only a very inefficient use of agricultural

[12] Stiglitz and Bilmes (2010) argue the opportunity costs probably add more: "For instance, many have wondered aloud whether, absent the Iraq invasion, we would still be stuck in Afghanistan. And this is not the only 'what if' worth contemplating. We might also ask: If not for the war in Iraq, would oil prices have risen so rapidly? Would the federal debt be so high? Would the economic crisis have been so severe?" Nine years after that update, the "forever wars" continue.

resources but it also contributes to a wide range of detrimental health outcomes—raising healthcare costs and reducing labor productivity.

For all these reasons, a comprehensive GND must include policies to address the pollution, deforestation, carbon emissions, and health implications of today's agricultural practices. However, it is difficult to assess the implications of such reforms for resource use. It is possible that natural resource exploitation would fall, but use of labor resources would rise if we shifted from huge factory farming to local, sustainable farming methods. It is possible that the net impact on labor demand could be positive, generating pressure on wages. This could be somewhat offset by lower demand for some kinds of resources (petroleum, pesticides, fertilizers) and lower costs of dealing with the negative impacts of factory farming. However, given the current uncertainties involved, we cannot provide an estimate of the impacts on resource use or on prices.[13]

3.6 *Other Sources of Inflationary Pressures*

Many fear that if the costs of medical care are shifted to the government that creates a large windfall for both employees and employers—who no longer must pay health insurance premiums—causing inflation. Employers would face substantially reduced labor costs as M4A replaces private insurance plans. For instance, in 2017, employers paid a median of about $13,000 in premiums per worker for an employer-based family coverage plan.[14] It is difficult to know how employers would react to the elimination of healthcare costs. If unions were strong, they might be able to induce employers to offer other kinds of benefits or to raise wages, which would increase incomes and thus consumption. Some have argued that employers would oppose the movement to M4A because they use

[13] See the following for discussion of some of the difficulties involved: Dmitrieva, K. 2019. "The Green New Deal Progressives Really Are Coming for Your Beef." Bloomberg. March 13. A summary of the best studies: https://ccafs.cgiar.org/bigfacts/#theme=food-emissions&subtheme=direct-agriculture; Gerber PJ, Steinfeld H, Henderson B, Mottet A, Opio C, Dijkman J, Falcucci A, Tempio G. 2013. Tackling climate change through livestock—A global assessment of emissions and mitigation opportunities. Rome: Food and Agriculture Organization of the United Nations. (Available from http://www.fao.org/docrep/018/i3437e/i3437e.pdf).

[14] In 2010, the median employer cost of healthcare was 12.8 percent of payroll (Claxton and Damico 2011). In 2017, employers paid about 70 percent of the cost of a plan for family coverage, while workers paid about 30 percent (KFF 2017).

good healthcare coverage as a recruiting device, so offering other kinds of benefits would be an option. We think that in the current environment, most of the savings will be retained by employers—although some might indeed offer longer vacations, shorter workweeks, or childcare as attractive benefits. Perhaps the savings would show up as profits—to be consumed by owners or highly paid executives, invested in new plant and equipment, or simply saved and perhaps used for stock buybacks that push equities prices ever higher.

If this windfall is spent, it would absorb resources, and hence our estimates of the resources released by M4A would be overstated. We think that is highly unlikely. When oil or other input prices rise, raising firms' costs, their spending on inputs rises and there is some pressure on inflation; when input prices fall, their spending on inputs falls. The pressure on aggregate prices is downward when costs of production fall. We suggest that the main effect of removing healthcare costs from employer costs will be deflationary, not inflationary. We believe US employers will use the savings to make their production more competitive, domestically and internationally, by lowering prices.

However, as workers' spending on healthcare premiums falls and take-home pay rises, some portion of this can go to increased consumption. Certainly workers will use some of that for increased consumption, but they can also reduce debt and increase saving. Still, there could be some increased consumer demand and some pressure on inflation.

4 Responding to Inflationary Pressures

4.1 Taxing the Rich

If we find that the resources released from destructive uses and mobilized are not sufficient to meet the requirements of the GND programs, inflation pressures will arise. It is difficult to know how large these will be, but if we do face inflation we need to have a plan for addressing it.[15]

Many advocate raising taxes on the rich to provide revenue to finance the GND. However, as we explained in this policy brief, the function of taxes is to release resources, rather than raise revenue. Hence, the type

[15] Note that we do assume that government will use other means to constrain prices, including cost controls in the single-payer system as well as prioritizing spending over the decade during which GND projects are phased in.

of taxes we use becomes an important consideration. For instance, taxes such as financial transaction taxes and offshore income taxes would have little benefit, since they may not be very effective in reducing spending and hence releasing resources for the GND.[16] Ironically, these kinds of taxes are often cited by GND proponents as particularly good sources of finance.

Taxing the rich can release resources if it reduces their spending on resource-using output.[17] Direct taxes on high incomes and wealth can also reduce inequality at the top. If resource use by the rich can be reduced, that will free up resources to be used in GND projects. Instead of building a third or fourth mansion for the rich, public housing for the poor could be provided. Instead of producing (and fueling) private jets, efficient forms of mass transit could be built and operated. The potential release of resources could be significant. However, to obtain these advantages would require high tax rates on both wealth and income—perhaps even higher than what Representative Alexandria Ocasio-Cortez has advocated (a top marginal tax rate of 70 percent). The political and technical barriers to imposition of sufficiently high tax rates would also be significant: the rich have the means and incentive to fight any move to raise their taxes, as well as to avoid tax payments if rates are raised.[18] We will not attempt to calculate how many resources might be made available because the uncertainty of passing tax hikes is too great. We will focus instead on alternative ways of combating inflation should it arise.

4.2 *Taxes on Wages—deferred Consumption*

Taxes on consumption and wages, on the other hand, are highly effective in lowering demand. But they will be neither politically popular nor especially fair, given the stagnant wages for the last 50 years. Workers have

[16] They might reduce high-speed trading and the incentives to move profits offshore—and to the degree that they are effective, tax revenues from those sources fall anyway, so cannot be counted as a financial source of revenue in any case.

[17] We are not arguing against higher income and wealth taxes on the rich, which can be used to reduce inequality and restore democratic governance. However, we cannot think of these as a source of "financial" resources for the GND and consequently tie our ability to implement the GND to the taxes paid by the rich.

[18] Dean Baker (2019b) makes a similar argument.

been "paying for" neoliberalism for two generations. We cannot ask them to pay more to reverse the damages done.

We thus propose a policy along the lines of Keynes's recommendation to "pay for" WWII, applying it to the GND "war" to save the planet. To reduce consumption of resources, we would need "targeted" taxes that would be effective in lowering aggregate demand. We need to reduce consumption demand by the less fortunate bottom 90 percent of Americans. However, this should be imposed in a progressive manner (with exemptions for the lowest-income earners), and any reduction of consumption should only be temporary and accompanied by a clear promise of a better standard of living later.

Senator Bernie Sanders has formulated a plan to "pay for" M4A, proposing a supplemental payroll tax on employers equal to an additional 6.2 percent of payroll, plus an additional payroll tax surcharge on employees of 2.2 percent, plus an array of other progressive taxes which he has projected would raise $1.4 trillion annually.[19] While we do not agree with his goal of raising revenue to "pay for" programs, we follow his lead in designing a method of reducing consumption to alleviate inflation pressure.

To allay fears of inflation, we propose to repurpose the payroll tax part of Senator Sanders' proposal. We would impose an average surcharge of 4.6 percent on the employee portion of the payroll tax. This could be applied in a progressive manner, with a lower surcharge on lower wages (and with the lowest wages exempt). The goal is not to raise funds to "pay for" benefits, but rather to defer compensation of employees—based on Keynes's plan for WWII. In this case, the purpose of the tax is to temporarily reduce workers' compensation. Over time, Social Security benefits would be raised to compensate workers for their sacrifice during the implementation of the GND projects that are required to reverse climate change. The surcharge payroll tax on workers could be phased out in the future as GND demands on resources decline and as the nation's productive capacity increases.

[19] He includes progressive income tax rates, taxing capital gains and dividends at the same rate as income from work, limiting tax deductions for the rich, adjusting the estate tax, and savings from health tax expenditures. His total tax take is estimated at nearly $1.4 trillion annually. We are not including the other tax changes here, nor are we adopting the tax increase to raise revenues to "pay for" M4A. The purpose is to release resources—in this particular case, it is to postpone consumption by deferring income.

As Keynes recommended, the deferred compensation (given as a boost to Social Security benefits) can be allocated in a progressive manner: with larger supplements paid to those currently receiving the lowest retirement benefits, tapering down for those earning the highest. Also, as Keynes recommended, the payroll surcharge (similar to his capital levy) would be imposed as soon as the GND begins, while the benefit supplements would be delayed until the inflation danger has passed. The surcharges paid by employees would also be phased out only after inflation pressures subside.

If we presume that these surcharges will be levied on the hospital insurance portion (HI) of taxable payroll (which, at about 45 percent of GDP, is 25 percent larger than the Old-Age, Survivors, and Disability Insurance [OASDI] payroll[20]), they would remove about $400 billion of income annually, equal to 2 percent of GDP.[21] We will add this as a net source of resources to offset GND requirements.

The surcharge on payroll taxes acts like a sinking fund—not to be used to "finance" future Social Security or the expansion of Medicare spending to all, but to reduce aggregate demand now on the promise that Social Security retirement will be boosted as deferred compensation. Note that workers are generally better off with the single-payer plan, even with the payroll tax surcharge.[22] Table 2 summarizes the net resource costs of the GND and the resources released by the payroll tax surcharge.

Lastly, we have not discussed imports as a source of resources to meet higher demand. We should not rely excessively on imports of the production needed for greening projects (solar panels, wind power equipment, electronic transportation, etc.), at least to the extent that other countries are mobilizing to use their resources to fight climate change themselves. However, as employment rises to boost general consumption, at least some of that will be met by imported consumer goods. This is not neces-

[20] The logic behind applying them to HI rather that to OASDI is not only that the base is larger but also that the tax is designed to offset some of the benefits of moving to a universal single payer.

[21] See SSA (2018) and CMS (2018).

[22] The median wage is about $32,000. The average annual premium for family coverage is over $19,000, of which about $13,000 is paid by the employer and $6000 is paid by the employee. By contrast, the payroll tax surcharge would be $1472 on the median worker's wage.

Table 2 Summary of net GND resource use (percent of GDP)[23]

Job guarantee	1.0
Greening Projects	5.0
Medicare for All	−3.7
Miscellaneous GND Projects[24]	0.0
End Forever Wars	−1.0
Tax the Rich	0.0
Payroll Tax Surcharge	−2.0
Net Increase in Resource Use (without payroll tax surcharge)	1.3
Net Increase in Resource Use (with payroll tax surcharge)	−0.7

Note Source of resources is negative; use of resources is positive

sarily something to be avoided, as many nations need to export consumer products to obtain dollar earnings they need to import green technology. This will help to attenuate inflation pressure—as it has done over the past two decades.[25]

5 Conclusions

With appropriate policies, Keynes thought we could "snatch from the exigency of war positive social improvements" (Keynes 1940, iii). The same thing is true today: we can either be reactive and respond to the calamities created by climate change as they happen (as we already do

[23] We acknowledge that some of the resources released by the fossil fuel and healthcare sectors may not be appropriate for GND projects, although a good case can be made that a lot of the workers would be able to contribute to either working in GND projects or helping with administration.

[24] Includes student debt relief and free college, public infrastructure, and universal childcare (some of which is included in the JG resource requirement). While many of the "miscellaneous" GND projects (additional public infrastructure, free public colleges, job training, childcare) will require resources, by increasing productivity they will also supply resources—so we have assumed their resource use nets to zero. While this should be true over the long run, there could be a net demand for resources in the early years. Whether this is inflationary depends on whether they are phased in as resources are made available; that, in turn, requires careful planning.

[25] This helps to explain why the Fair model's simulation of the JG program (see Wray et al. 2018)—which boosted employment by 19 million, raised wages to or above $15 per hour, and increased annual GDP by half a trillion dollars—projected almost no inflation pressure.

with droughts and hurricanes), or we can use this crisis as an opening for progressive change. This is why the GND includes a wide range of social initiatives: jobs for all, ending forever wars, taxing the rich, student debt relief, free public colleges, access to child and elder care, and so on. At the same time, we need to plan for the implementation of these GND components at a pace that does not significantly raise inflation. What Keynes was arguing against in *How to Pay for the War* was the policy of allowing the economy to adjust to the realities of the war through an "inflation tax." If we do have to fight inflation, we need to ensure the battle is not waged on the backs of workers. The inflation tax permanently lowers consumption; deferred compensation only postpones consumption.

Our proposal provides resource numbers for each category: resources released, resources needed, and net increase of demand for resources (see also Nersisyan and Wray 2019). Our plan would be approximately net zero in terms of resource use if we impose the payroll tax surcharge. We do not believe that an increase in demand equal to 1.3 percent of GDP (our projection in the absence of a payroll tax surcharge) would cause significant inflation anyway—so even without the payroll tax surcharge, we do not anticipate inflationary pressures. However, we leave that as an option, and of course the surcharge could be higher or lower as necessary. Even with the surcharge, households would be better off in the short run—and substantially better off in the long run as their retirement pay is boosted.

Our main goal has been to set out a framework for analyzing the true cost of the GND—not to promote any particular estimate. As we noted above, there are considerable uncertainties associated with estimating the resource needs and savings from major economic programs. Our goal is to change the debate from estimating financial costs to careful assessment of resource needs and availability. We need an informed discussion of the best method of reducing resource use—should that become necessary—so as to free up resources for the GND. We have discussed deferred compensation as a preferred method. However, we believe that if the requirements turn out to be much larger than what we have estimated, we can also explore the other methods that were successfully used in WWII: patriotic saving (which is voluntary deferred consumption), price controls, rationing, and additional taxes. Most importantly, if taxes are to be used, they must be formulated to reduce resource use—not to "raise revenue."

References

Baker, D. 2019a. Testimony by Dr. Dean Baker, Senior Economist, Center for Economic and Policy Research, US House of Representatives Rules Committee, Re: HR 1384—Medicare for All Act of 2019a. April 30.

Baker, D. 2019b. "MMT and Taxing the Rich." Beat the Press. Washington, DC: Center for Economic and Policy Research. February 15.

BEA (Bureau of Economic Analysis). 2019. "National Income and Product Accounts. Table 5.3.5. Private Fixed Investment by Type." Last accessed April 30, 2019.

Bernie 2020. n.d. "The Green New Deal." Accessed on December 21, 2019. Available at: https://berniesanders.com/en/issues/green-new-deal/.

Bruenig, M. 2019. "US Workers Are Paying High Taxes. But Without Any of the Benefits." *Jacobin*. April 14.

Claxton, G., and A. Damico. 2011. "Snapshots: Employer Health Insurance Costs and Worker Compensation." San Francisco, CA: Kaiser Family Foundation. February 27.

CMS (Centers for Medicare and Medicaid Services). 2018. Table V.F1: Annual Revenues and Expenditures for Medicare and Social Security Trust Funds and the Total Federal Budget, Fiscal Year 2017. Published in "2018 Annual Report of the Boards of Trustees of the Federal Hospital Insurance and Federal Supplementary Medical Insurance Trust Funds." Washington, DC: CMS.

Dmitrieva, K. 2019. "The Green New Deal Progressives Really Are Coming for Your Beef." Bloomberg. March 13.

Heal, G. 2017. "Reflections—What Would It Take to Reduce US Greenhouse Gas Emissions 80 Percent by 2050?" *Review of Environmental Economics and Policy* 11(2): 319–35.

Holtz-Eakin, D., D. Bosch, B. Gitis, D. Goldbeck, P. Rossetti. 2019. "The Green New Deal: Scope, Scale, and Implications." Washington, DC: American Action Forum. February 25.

Jacobson, M. Z., M. A. Delucchi, Z. A. F. Bauer, S. C. Goodman, W. E. Chapman, M. A. Cameron, C. Bozonnat, L. Chobadi, H. A. Clonts, P. Enevoldsen, J. R. Erwin, S. N. Fobi, O. K. Goldstrom, E. M. Hennessy, J. Liu, J. Lo, C. B. Meyer, S. B. Morris, K. R. Moy, P. L. O'Neill, I. Petkov, S. Redfern, R. Schucker, M. A. Sontag., J. Wang, E. Weiner, and A. S. Yachanin. 2017. "100% Clean and Renewable Wind, Water, and Sunlight All-Sector Energy Roadmaps for 139 Countries of the World." *Joule* 1(1): 108–21.

Keynes, J. M. 1940. "How to Pay for the War: A Radical Plan for the Chancellor of the Exchequer." London: Macmillan.

KFF (Kaiser Family Foundation). 2017. Figure A: Average Annual Firm and Worker Premium Contributions and Total Premiums for Covered Workers for Single and Family Coverage, by Plan Type, 2017. Published in "Employer

Health Benefits: 2017 Summary of Findings." San Francisco, CA: Kaiser Family Foundation. September 19.

Li, Y. 2019. "The Fed Chairman Says the Relationship between Inflation and Unemployment Is Gone." http://cnbc.com/. July 11.

Nersisyan, Y., and L. Randall Wray. 2019. "How to Pay for the Green New Deal." Working Paper No. 931. Annandale-on-Hudson, NY: Levy Economics Institute of Bard College. May.

OECD (Organisation for Economic Co-operation and Development). 2019. Health Spending (Indicator). OECD Data. https://doi.org/10.1787/8643de7e-en. Last accessed May 5, 2019.

Pollin, R., H. Garrett-Peltier, J. Heintz, and B. Hendricks. 2014. "Green Growth: A U.S. Program for Controlling Climate Change and Expanding Job Opportunities." Center for American Progress and Political Economy Research Institute. September 18.

Pollin, R., J. Heintz, P. Arno, J. Wicks-Lim, M. Ash. 2018. "Economic Analysis of Medicare for All." Amherst, MA: Political Economy Research Institute. November 30.

Sanders, B., and M. Lee. 2019. "Congress Just Told Trump to Get US Troops out of Yemen. Next, Afghanistan?" USA Today. April 4.

Shane, L. 2019. "A Plan to End Afghanistan War: Declare Victory and Give $2,500 Bonuses to Vets." Military Times. March 5.

SSA (Social Security Administration). 2018. "The 2018 Annual Report of the Board of Trustees of the Federal Old-Age and Survivors Insurance and Federal Disability Insurance Trust Funds." Washington, DC: SSA.

Stiglitz, J. E., and L. J. Bilmes. 2010. "The True Cost of the Iraq War: $3 Trillion and Beyond." Washington Post. September 5.

Tcherneva, P. 2018. "The Job Guarantee: Design, Jobs, and Implementation." Working Paper No. 902. Annandale-on-Hudson, NY: Levy Economics Institute of Bard College. April.

Wray, L. R., F. Dantas, S. Fullwiler, P. R. Tcherneva, S. A. Kelton. 2018. "Public Service Employment: A Path to Full Employment." Research Project Report. Annandale-on-Hudson, NY: Levy Economics Institute of Bard College. April.

Wray, L. R. 2019. "MMT Responds to Brad DeLong's Challenge." New Economic Perspectives. March 12.

Wray, L. R. 2015. "Modern Money Theory: A Primer on Macroeconomics for Sovereign Monetary Systems." London: Palgrave Macmillan.

Sustainability and Social Policies

Sustainability and Social Policy Nexus

Mahmood Messkoub

1 Introduction

This paper is concerned with the macro-level sustainability of social policies, rather than sustainability of specific social policies like health and education in terms of their environmental impacts, important as they are. At macro-level a sustainable social policy is intrinsically linked to a sustainable economy and society. Considering the emergency nature of climate change based on scientific evidence on global warming of 1.5 C above pre-industrial levels (IPCC 2018), sustainability of life on earth requires a paradigm shift in our approach to how the economy and society should be organised and managed.

Special thanks are due to Ray Bush and Jeremy Leaman for their insightful and valuable comments on an earlier draft of this paper. All the remaining errors are, alas, mine!

M. Messkoub (✉)
International Institute of Social Studies (ISS, the Hague), Erasmus University of Rotterdam, Rotterdam, The Netherlands
e-mail: messkoub@iss.nl

N. Karagiannis and J. E. King (eds.), *Visions and Strategies for a Sustainable Economy*, Global Institute for Sustainable Prosperity,
https://doi.org/10.1007/978-3-031-06493-7_8

For decades economic growth has been one of the most important policy objectives in the capitalist market as well as centrally planned economies in order to deliver on human development targets and reduce poverty and inequality at national and international levels. Redistributive social policies played a crucial role in mobilising resources generated by economic growth and prosperity to improve human welfare measured, e.g., by low infant and child mortality and higher life expectancy. Moreover, social policies such as state support to children and families, minimum wage, unemployment social insurance and pensions made important contributions to the welfare of the population through different stages of life that have come to be known as a welfare state (Pierson 1998; Pierson et al. 2014).

Reliance on growth has not been without its costs in terms of depletion of environmental resources and pollution. The early industrialising developers of the North have been on the forefront of this destruction of environment both at home and abroad through centuries of colonial and imperialist resource extraction. This is the environmental debt of the high-income countries to the rest of the world (see, e.g., Patel and Moore 2017; Redclift 2018; Hickel 2020; Ajl 2021). The political/political-economy ecology literature has raised important questions about a growth-centred economic strategy that has led to a rich literature on whether or not economic growth is needed for improving human welfare (see, e.g., ibid., Berr 2015; Jackson and Victor 2016).

It is in this context of economic growth and development that this paper addresses social policy issues and their contribution to improving human welfare and achieving a sustainable society. It begins with a critique of conservative environmentalism that is inspired by Malthusian population pressure (with all its social policy implications) that to some extent also informs the degrowth approach. It would then ask how to meet the increasing health, education and other social needs whilst minimising the depletion of natural resources. The paper would argue that the answer to the question of a sustainable social policy in part lies in an economic model, a la Kalecki and others, that can manage/negotiate the composition of output whilst investing in resources to reduce depletion of natural resources and greenhouse emissions.

This is a growth strategy based on 'the human theory of needs' that meets the needs of current generation and provides some measure of inter-generational justice. The welfare and social policy counterpart of this must, the paper will argue, involve public and collective provisioning of socially necessary services of health and education as well as a range of other care services that will reduce per capita cost through economies of scale and scope whilst providing an equitable access to these services—universal provision and access and not targeting is at the heart of this approach.

To deliver on an eco-development strategy that addresses the social needs of any society this paper pays particular attention to the relationship between an active civil society and an activist democratic state with its concomitant role of regulating the market economy (Berr 2015).

2 Social Policy and the Question of Sustainability

Sustainability is 'defined as development that meets the needs of the present without compromising the ability of future generations to meet their own needs' (WCED 1987, p. 40) sets out a boundary on current economic growth and consumption since earth resources are limited. This is the first condition of sustainability. Besides limits on resources, sustainability also has to take into account of pollution and absorption capacity of the earth (Fitzpatrick 2011). We might consider this as the second condition of sustainability. These are the minimum conditions that any sustainable social policy has to fulfil.

The relationship between social policy and its environmental impacts in modern political-economy literature dates back to the work of Malthus who in the late eighteenth century articulated one of the earliest, and undoubtedly most influential, relationships between needs and activities of people and their impact on limited land and natural resources (Malthus 1798). In a nutshell he argued that given the limited supply of land and the limits to the productivity of that land, the rate of growth of population would surpass the rate of growth of food production because of the diminishing returns to land as the number of people increase. Malthus' important assumption was that food production would grow at an arithmetic rate whilst population grows at a geometric rate, thus food production per head would decline steadily. To counter this development, he envisaged positive and negative checks on population, the former resulting from disease and famine and the latter from reduction in births due to delayed marriage and in general less sexual activity.

Poverty in the world of Malthus was due to ignorance and indolence of the poor and the solution was keeping their number low.[1] It is the change in the age of marriage and lowering of fertility which are at the heart of a Malthusian solution to negative impact of population on the environment. Population control as a social policy has been the cornerstone of some influential environmentalist writers in the twentieth century.[2] A similar Malthusian argument has been made against international migration for putting pressure on resources of host, especially the high-income, countries. The argument has been based not only on the absolute increase of population but that the consumption pattern of migrants matches the high consumption levels of host countries. Combination of the two effects, it is argued, inevitably leads to more depletion of earth resources, especially in the destination country (Angus et al. 2011, ch. 9; Esmailian 2015; Weber and Dabbs Sciubba, 2019).

Empirical evidence on fertility and food production does not support Malthus' assumption and its consequences. Not only has world-wide fertility been falling, but some developing countries have had the fastest decline in fertility in history (Mehri et al. 2020; Messkoub 1999; Thirlwall and Pacheco-Lopez 2017). Besides, growth of food production has outpaced population growth in the past 50 years (FAO 2002) and is predicted to continue (albeit at a slower rate) well into the middle of the twenty-first century (FAO 2009) with no threat to world food supply and market in the short-term (FAO 2021).

Other classical economists like Ricardo and John Stuart Mill did not fully share the pessimism of Malthus as far as the negative impact of population was concerned. However, it was left to Karl Marx to argue that the problem of poverty was not the pressure of population in relation to the means of consumption, as in Malthus, but pressure of population against the means of production. According to Marx, there is a constant pressure to increase surplus value by increasing the productivity of labour; thus the growth of accumulated surplus, if invested, leads to increased productive capacity. But this increased productive capacity is at the expense of increased population because of labour-saving machinery. Demand for labour will not match the supply of labour as population

[1] For a detailed critique of Malthus see Foster (1998) and references therein.

[2] See, e.g., Ehrlich and Ehrlich (1968, 1970), Hardin (1968). For a critique and further references see Angus et al. (2011) and Thirlwall and Pacheco-Lopez (2017) Chapters 4 and 11 and references therein.

increases leading to unemployment and creation of a reserve army of labour which presses down on wages. Poverty is not then due to too many people pressing over too few resources—Malthus' overpopulation thesis—but unequal distribution of productive resources (including land) that leaves too many people without the resources to earn a living or to produce for themselves.

The two approaches to the question of population and poverty lead to two different policies on poverty alleviation and improving welfare. In Malthus' view the original Poor Law of 1500 and its reformed version in the nineteenth century made the poor dependent on charity. Moreover, he argued, any support to the poor would be at the expense of the rest of the population because the means of subsistence are limited and fixed. He further argued that if a fixed amount of wage goods were to be distributed among a growing population or paid as higher wages to the working population, the price of wage goods would increase. The resulting wage inflation would erode the purchasing power of higher wages. He concluded that those who do not contribute to production and earn their living do not deserve to be supported. In short, any charity to the poor or increase in wages would be detrimental to the welfare of workers and the rest of the population.[3]

For Marx, on the other hand, poverty was a necessary outcome of the exploitation of the workers (their receiving only a part of the surplus value they produced) and their lack of access to the means of production. This was an issue of distribution of output as well as access and control over the means of production, including land and other assets. The distribution of output and resources have been driven by class and social struggles to improve living conditions. That in turn has put the question of distribution firmly on the agenda of social reformers since the middle of the nineteenth century, which also has come to define the role of the state in the economy and society.

With this brief introduction on the origin of the relationship between environmental issues and economics, we can now turn to a discussion of the evolving relationship between economics and social policy in the twentieth century.

[3] For a more detailed discussion of Malthus' view on population and 'wage-fund' see Foster (1998). See also Rezai and Stagl (2016) on the historical linkages between environmental concerns and the development of economics.

3 From Normative to Positive Social Policy: Distributional Issues

Social policy that emerged from the above discussion on the origin of the environment and social policy nexus was a normative one. Malthusians solution for overpopulation and Marxian argument for distribution were more ideas than policies. Political struggles in the eighteenth and nineteenth centuries against absolutism and arbitrary rule and emergence of nation-state turned normative social policy ideas into positive social policies. Some of these social policies like universal education, public health measures and the prohibition of child labour were part of the agenda of capitalist industrialisation that needed an educated and healthy labour force. Others like state pension, public healthcare, child support and unemployment benefit are the result of working people's struggle for a better life and a bigger share of the national output and resources. These are distributional issues that any economy has to address subject to sustainability conditions. But first let us provide a brief outline of what a positive social policy is.

3.1 *Positive Social Policy*[4]

A positive social policy is 'collective intervention in the economy to influence the access to and the incidence of adequate and secure livelihoods and income' (Mkandawire 2004, p. 1). This provides a good starting point to locate the role of social policy in a modern market economy with its unequal distribution of resources and insecure livelihoods. It is important to emphasise the broad objectives of social policy with regard to redistribution, protection and transformation (ibid.). This is in contrast to the narrow view of social policy which limits its brief to matters of health and education and to a 'safety net' and social protection of those people in social and economic distress due to, e.g., unemployment, poverty, disability or old age.

We need to distinguish between social welfare and social policy. Whilst social welfare refers to any kind of mutual help and socially organised activities—whether provided by the state or non-state actors—social policy is state-led and state-organised and aimed at supporting individuals

[4] This section draws on the author's earlier works: Jawad, Jones and Messkoub (2019) and Messkoub (2021).

and households and forms part of the broad economic and development agenda of diversification of an economy that requires a healthy and educated labour force and a stable social environment. In this regard social policy can be viewed as a social contract between labour and capital—is citizens and the state—that are important in the growth and development process of a nation state (Briggs 1961; Ghosh 2011).

Historically, societies have witnessed a move from small-scale, largely faith-based or mutual support networks among people sharing similar lineages, works and cultures, to large-scale state-sponsored social support systems—a transition that has tracked closely the emergence of nation states. Faith-based social support has usually been founded on charitable principles of 'doing good' and helping people in need and alleviating poverty, whilst mutual support networks have been founded on principles of self-help and reciprocity.[5] These developments are also reflected in the structure and time path of social welfare-provisioning around the world where faith-based and mutual support charitable social services are combined with extensive services provided by the state (Messkoub 2021).

The distinction between social welfare and social policy would also draw attention to the role of non-state social actors in welfare-provisioning. Examples of such social institutions are households caring for their members and non-state secular and religious mutual charitable organisations that provide health and education services.

There is also an important analytical difference between social provisioning organised by the state as part of its social policy agenda that is based on the principles of 'rights, regulation and redistribution' in the context of a social contract, and social provisioning provided by non-state actors that are based on 'reciprocity, inclusion and trust' as well as altruism. The premise of 'rights' has an important implication for citizenship, entitlement, claim and access to various services as well as legal and regulatory provisions of social policies that also provides the backdrop for political aspirations and claims on the state.

Such has been the history of the welfare state in Europe and the development of state-organised welfare-provisioning around the world. The history of the Poor Law in the nineteenth century Britain and the fact that the Poor were not considered full citizens with rights is a testimony

[5] For a brief discussion of history of social welfare, see Briggs (1961) and Marshall (1950).

to the importance of rights in the development of the welfare state. As T. H. Marshall (1950) stated:

> The Poor Law treated the claims of the poor, not as an integral part of the rights of the citizens, but as an alternative to them—as claims which could be met only if the claimants ceased to be citizens in any true sense of the world. For paupers forfeited in practice the civil rights of personal liberty, by internment in the workhouse, and they forfeited by law any political rights they might possess. This disability of defranchisement remained in being until 1918... The stigma which clung to poor relief expressed the deep feelings of a people who understood that those who accepted relief must cross the road that separated the community of citizens from the outcast company of the destitute. (p. 31)

Rights also influence and contribute to what Fraser (1989) has called 'the political imaginary of social welfare', an imaginary that may have originated in political promises on welfare-provisioning. But rights put them on a firmer footing which, combined with regulations of, e.g., health and safety at work, wages and labour relations would open up other areas to claim a fairer share of the fruits of one's labour from private and public employers (Pierson 1998, ch. 4).

In any welfare regime, the relative weight of different institutions (state, market, households and non-state organisations) critically depends on the balance of political forces and economic conditions in a country that over time sets the trajectory, achievements and vicissitude of social policies (Esping-Anderson 1990; Kwon et al. 2009). Herein lies the path dependency of social policy and welfare regimes in the world.

This broad view of social policy is closely associated with the emergence of welfare state and Keynesian post-WWII economic policies (Sawyer 1989) that also found its counterpart in the state-centred development in centrally planned economies (Pierson 1998).

Distributional concerns are at the core of any economic theory concerned with the welfare of the population in the present and in the future. Whilst in the short-run distributional issues are more concerned with the current production level than increasing production capacity through investment, in the medium- to long-run increased production and investment issues dominate since increasing welfare, especially in countries which start from a low level of welfare of the population, require an expanded production. Even in countries which have achieved a high level of human development (in terms of low of infant and child mortality, high life expectancy, high levels of literacy and educational attainment by both genders, high level of gender equity in law and practice, etc., and even high level of 'happiness' as in Denmark) need to increase production in order to replenish depreciating capital stock and to satisfy increasing demand as incomes rise and people live longer.

3.2 *Growth, Sustainable Development and Social Policy*

The growth trajectory of an economy and the composition of output of that growth provide the critical link between social policy and sustainable development. Since the early phases of globalisation in the sixteenth century and especially since the industrial revolution in the eighteenth century growing economies have relied on a range of resources that have been shipped around the world thus depleting the exhaustible resources of the earth. Earlier phases of globalisation were characterised by a smaller world population, less destructive technologies and slower means of transport and communications compared with the current phase which gives sustainability a much more globalised character. Yet, the earlier phase—the era of colonial expansion and control—which was based on the production and extraction of raw materials using indigenous as well as slave/indentured labour did lead to large-scale destruction of nature in the colonised world. The earlier phase of globalisation has left a legacy of environmental degradation, and production and trade patterns to low- and middle-income countries that has hampered their development over time. It is in this context that the question of the environmental debt of the high-income to other countries arises. Compensation for such environmental debt has to be factored into any global sustainability strategy. Let us note that in a globalised world sustainability is borderless.

A sustainable growth is at the heart of a sustainable social policy. Nation states promote growth, as measured by GDP, for its own sake as an indicator of the prosperity of a country, as well as for its importance for the state's finances for funding various social policies. But the very notion of sustainable growth has been questioned in relation to exhaustible national and international resources giving rise to a growing literature on degrowth. It is at this crossroads of growth and degrowth that we need to discuss sustainable social policies (Fitzpatrick 1998; Hickel 2019, 2020; Patel and Moore 2017). The composition of output is an important issue for a sustainable economy; a composition which contrasts to that of a capitalist economy dominated as it is by effective demand and the so called 'consumer sovereignty'. In this context the question of 'What to produce?' cannot be separated from the question of 'For whom to produce?', with all its distributional implications. We need to ask some very fundamental questions about 'What?' the economy should produce and 'For whom?' and for 'What purpose?' and 'How?'.

The question of 'needs' versus 'wants' is at the centre of this debate. Gough (2015) has argued that sustainable welfare should be based on a theory of human needs rather than consumer preference and sovereignty. 'Preference satisfaction [of "wants" or "needs"] cannot provide a logical, ethical or practical conception and measure of human well-being—and especially so when we must consider well-being on a global and inter-generational scale' (Gough 2015, p. 1194). Because current preferences are subjective (e.g., preferring Mozart to Beethoven) and preferences of the future generation cannot be revealed. That is why 'needs' (e.g., for water and certain amount of daily calorie-intake) which are objectives and can be socially identified, quantified and universalised are better for organising production, consumption (as needs satisfiers) and distribution. 'Need' in the context of the theory of human needs:

> ...refers to a particular category of goals which are believed to be universalisable. The contrast with wants—goals that derive from an individual's particular preferences and cultural environment—is central to our argument. The universality of need rests on the belief that if needs are not satisfied then serious harm of some objective kind will result... we define serious harm as significantly impaired pursuit of goals which are deemed to be of value by individuals. (Gough 2015, pp. 1194–1195)

The most basic of needs is physical health that has a two-way relationship with the important facets of human autonomy: mental health, cognitive understanding and opportunities for social participation. To satisfy the basic need of health, the theory of human needs stipulates the following universal requirements: adequate food, nutrition, water and shelter; non-hazardous work and physical environment; physical and economic security; safe birth control and childbearing and appropriate health care.[6]

[6] For a full outline of these linkages see Gough (2015, p. 1196). See also Stewart (2019). For a well-argued philosophical approach to the concept of need and its practical application to environment based on precautionary principle see Wiggins (2005).

It is obvious that these requirements come in direct conflict with environmental degradation and pollution as well as climate change that would seriously harm and impair the ability of human beings to meet their most basic needs of health. That is why directing and managing production, consumption and distribution of resources of an economy is essential for a sustainable economy and a sustainable social policy. Or as Gough (2015) states:

> ...the goal of securing well-being in the face of climate change has resulted in two ethical demands: sustainability (in the form of the emissions envelope) and equality (in the form of prioritising the satisfaction of human needs, present and future, over surplus wants). In theory, the needs and sustainability criterion must continuously be iterated until they become compatible. (p. 1208)

The socio-economic context of a growth/degrowth strategy also matters which raises an important methodological issue: is a sustainable environment a private or public good? Eckersley (2004) has argued that the sustainable environment cannot be treated as a private good because of the negative externalities of private decisions with regard to investment, production and consumption. Such ecological costs are a public 'bad' that 'restricts the ability of environmental victims (both inside and outside the borders of the nation state) to enjoy the full range of freedoms that liberalism supposedly upholds, including the freedom to participate or otherwise be represented in the making of decisions that bear upon their own lives' (Eckersley 2004, p. 242).

Economic theories which are based on the agency of individuals with regard to production and consumption decisions have treated sustainable environment as a private good that can be produced and consumed with the right incentives. Market-based solutions to achieve environmental sustainability are a logical and practical corollary of such a methodology. But the very nature of environmental problems—negative impact of individual decisions on others and the interrelatedness of these actions—calls for treating them as a public and not private matter and therefore regarding a sustainable environment as a public good. It is in this context that the role of state and in general collective action in environmental protection and sustainability at national and international levels becomes crucial. The fiscal and regulatory roles of the state should be used to protect the integrity of the ecosystem, and as importantly, 'to redistribute resources and otherwise influence life opportunities to ensure that the move towards a more sustainable society is not a socially regressive one—a very real prospect if environmental goals are not properly integrated with social justice goals' (Eckersley 2004, p. 12).

Let us focus on some key issues on the role of the state with regard to the nexus of sustainability and social policy. The regulatory role of state in environmental matters must be combined with its fiscal measures to finance social policies which protect individuals and households and promote their capabilities overtime in order to ensure that environmental protection is progressive by reducing inequality in the society. We return to these issues in our discussion of policy choices in the context of economic growth and the degrowth agenda, but first let us turn to some of the key questions for a sustainable social policy.

3.3 *The Key Questions for a Sustainable Social Policy*

Fitzpatrick (2011) has argued that sustainable social policies have to address the economics of sustainable growth by combining zero growth and market-based policies of carbon quotas/taxes with distributional policies at national and international level. Distributional policies in the context of carbon pollution should take note of higher per capita contributions to pollution of the richer people in any society and higher pollution per capita of the rich high-income countries that suggests national and international costs have to be borne mainly by the rich. Based on works by Stern (2009) and other sources, Fitzpatrick suggests that a redistributive fund of 5% of GDP of high-income countries could go some way to reducing poverty internationally and pay for adaptive policies to meet environmental challenges and help with sustainability in low and middle income countries. This redistributive fund is part of the policy agenda emerging from the Environmental Kuznets' Curve (EKC)-an inverted U shaped curve that relates, as a stylised fact, income per capita to environmental pollution. EKC stipulates that in the process of development initially pollution increases as income per capita increases, but it decreases after certain level of per capita has been reached (Dinda 2004; Environment and Development Economics 1998).

The question of social justice and redistribution goes beyond reducing inequality among individuals, households and countries. There is a temporal dimension to social justice that involves inter-generational issues as well as the dynamics of equity. As regards the latter, distributional policies should have a life-long perspective of a generation. Welfare of an individual is a continuous process that starts before birth. The inter-generational issues take us back to the basic question of sustainability, which is about the welfare of the future, including the unborn, generation who should have access to the same environmental resources as the current generation. The two generations, however, have substantial overlaps in terms of time and space and separating them by the labels of 'current' and 'future' generations in theoretical discussions and analytical modelling should be complemented with empirical work based on socio-economic data and analysis in order to formulate sustainable social policies.

A green social policy should also involve and engage the population and communities at every level in order to combine a state-centred social policy planning and budgeting with grass root involvement in policy debate, design and implementation (Fitzpatrick 2011, ch. 3).

A range of other issues has also been put on the agenda of social policies that are about sustainability of specific social policy projects like the greening of health or education infrastructure or micro-level personal carbon trading (ibid.). These issues are part and parcel of any debate on radical sustainable social policy but, important as they are, they go beyond our macro approach. Common to all these micro- and meso-level issues is how to achieve sustainability whilst minimising the impact on the lower income groups with a combination of direct provisioning and preferential taxes and subsidies to individuals and households.

Sustainability also has to take note of the impact of demographic change that has important implications for the commitments of the welfare state overtime. All private or public pension systems enter into a legal and social contract with the working people according to which working people make regular contributions to a funded or pay-as-you-go pension system on the promise of a pension when they retire. A public state pension system based on the right of citizens to a pension is effective when citizens reach the landmark age of retirement. This means right to a pension effectively matures at retirement. Similar arguments hold for the right to medical care, social care, etc., at retirement. To fulfil these rights states have to plan for a future based on the managed growth of national economy and distribution of national income, which are critical challenges when environmental concerns about growth have to be reconciled with the extra resources needed to manage demographic change.

4 Growth or Degrowth?

A sustainable social policy at macro-level critically revolves around economic growth. The trajectory of human development around the world closely follows the trajectory of expanded production and economic growth.[7] However, the achievement of human development goals in different countries has depended on how the benefits of growth have been distributed, directly through social policy measures such as provision of health and education and indirectly by raising the capabilities of population and by giving them access to resources.

A sustainable social policy is then about sustainable growth and distributional issues related to it. A sustainable growth policy covers the spectrum of zero (or even negative) growth (degrowth) to moderate growth, both having the objective of preventing further depletion of the planet's exhaustible resources, combined with the reduction of greenhouse gas emissions. Moreover, in the degrowth literature degrowth 'indicates the removal, descent, and reversal of economic growth' … and 'is operationalized by the notion of matter/energy throughput, which measures the biophysical metabolism of an economy, that is, how much nature humans use and transform in absolute terms. An economy is degrowing when its matter/energy throughput reduces' (Robra and Heikkurinen 2019, p. 1). This raises fundamental issues on the limits to human activities and how to meet human needs.

Degrowth literature also questions the validity of a growth-based welfare state on the basis of the 'human and social costs of economic growth—stress-related illness, the diseases of affluence, unsatisfying labour and the direct economic costs of social ills—[making] the "economy of the welfare state" self-defeating' (Pierson 1989, p. 184).[8] If we also consider social movements built around the degrowth, which has raised important questions on the link between human well-being and prosperity on the one hand and increased production and consumption on the other, we have strong theoretical and social reasons to question economic and social systems predicated on ever-expanding production and consumption. It is in this context that 'a counter-hegemonic

[7] For a good historical account of the importance of economic growth for the development of social policies and welfare state see Pierson (1989), ch. 4.

[8] For further discussion of the human cost of capitalist growth strategies see also Patel and Moore (2017) and Hickel (2020).

discourse to capitalism is ... called for' (Robra and Heikkurinen 2019, p. 9). That is in contrast to market-based approach of those in the Green movement (e.g., Pearce et al. 1989; Pearce 1992) or environmental economists like Nordhaus (2008) who rely on the efficiency of market solutions for resource allocation. It is important to note that 'consumer sovereignty' is one of the fundamental core assumptions of any market-based approach to environmental management that may well come into conflict with citizen sovereignty and social citizenship that are foundations of distributive and progressive social policies of modern states.[9]

The degrowth literature draws our attention to the limits, perhaps in the long-run, of a sustainable social policy, that takes us back to the issue of how to make growth sustainable in the short- to medium-run whilst meeting the goals of social policy. Some fundamental questions, however, still remain. In the debate on sustainability, social policies could no longer be based on the redistribution of a given level of GDP; the composition of GDP matters for social policy. Moreover, negative environmental impacts of economic growth on the health and welfare of population should not be treated as simply a matter for compensatory social policies such as free hospital care in areas affected by noxious industrial emissions, or relocation of population with compensation for the lost land in order to, e.g., build a dam. Other important issues are related to the international implications of national growth that concern the international externalities of national economic decisions (e.g., polluting rivers that cross borders and global warming) and differences in growth and resource-use of national economies. The latter should take into account low- and middle-income countries' need to grow to meet their economic and human development challenges.

Finally, there is the all-important issue of financing national social policies. Could nation states rely on national tax capacities in the face of the need to increase social spending on an ageing population, the increasing cost of medical technologies and other social needs? Tax capacity critically depends on rate of economic growth that in turn raises the question of compatibility of a sustainable social policy with growth.

The next section will discuss some of these macro-level concerns of a sustainable social policy.

[9] For further discussion on the importance of continuing role of state in implementing a progressive green agenda see Pierson (1989), Conclusion, pp. 201–208.

4.1 *What Kind of Growth?*

Is growth necessary for a reduction in inequality and improved living conditions and welfare of population? The answer to this question is 'no', because at zero growth a given level of output could be distributed more equitably and economic resources could be used more appropriately (whilst reducing waste) to achieve higher welfare. However, increased population as well as increased needs and demands especially in societies that start from a low level of human and economic development require some increase in national output. What concerns us here is the character—in terms of technology and composition of output produced—of this growth and the distribution of that output and broader distributional issues. Let us recall that an important objective of progressive social policies is to reduce inequalities of income and wealth. It is in this context that we need to discuss limits to growth.

The question of limits to growth and resource-use in orthodox approaches to macroeconomics is simply about providing the correct price signals to the market, combined with appropriate taxes and 'cap and/or trade', and regulation policies on energy use and greenhouse emissions. It is argued that a supply-side approach, based on rational behaviour and perfect markets, directs resources towards energy and resource savings technologies. This is challenged by ecological and post-Keynesian economists who have argued that not only growth has to be based on higher levels of resource productivity but waste has to be minimised to reduce or keep throughput constant. Moreover, these economists argue that low-growth should be treated as a policy objective which requires directing the volume and composition of investment using regulatory power of the state as well as harnessing financial resources for sustainable investment (Rezai and Stagl 2016).

We also need to consider the relationship between growth, distribution and inequality. Piketty (2014) argued that lower economic growth in the long-run would increase inequality in favour of capital; as the share of capital in national income would increase. The key to the increasing share of capital is the substitution of capital for labour to increase the productivity of labour and increase the rate of exploitation and increase the rate of profit and hence increase the share of capital in total output. Tight labour markets also contribute to the substitution of capital for labour. As a result, and over time, capitalists would accumulate more wealth. In short, low economic growth aggravates both inequality of income and wealth, which however can be corrected with appropriate redistribution policies using progressive taxation (Jackson and Victor 2016). But this crucially depends on the relative strength and influence of workers in the design and implementation of tax and spend policies. Moreover, a degrowth strategy should also deal with the distribution of wealth to reduce inequality.

In orthodox market-based sustainable growth strategies environmental mitigation measures such as carbon taxes and cap and/or trade schemes play an important role to incentivise less energy use through the price mechanism. By raising the cost of energy, whether directly or indirectly, these measures have been found to have regressive distributional impacts in high-income countries where the share of expenditure on energy for low-income households is larger than that for high-income households. This calls for compensation measures such as the introduction of differential energy prices and targeting support to services, such as public transport, which are used more intensively by lower income groups. It has been argued that in low-income countries energy is a 'luxury good' for the poorer households, and therefore taxing energy use would be progressive (Buchs et al. 2011).

Notwithstanding the contribution of progressive mitigation and compensation policies to a sustainable social policy agenda in the short-run, what matters most in the long-run is a progressive low-growth or degrowth strategy which commits us collectively to focus on the composition of output following the theory of human needs. In this regard Kalecki's writing on the composition of output during the course of development in low-income countries provides important insights (Kalecki 1976; Stewart 1985; Fitzgerald 1990; Ghose 2011). Kalecki was not directly concerned with environmental and sustainability issues but how to manage the competing demands of an economy for consumption goods and investment goods, with the latter being crucial for growth. He argued that the production of consumption goods (wage goods) should be at a level to ensure the well-being of population and allow sufficient resources to be diverted to investment goods. Note that we distinguish between 'essential' and 'non-essential' consumption goods; however, the essentials are not limited to, e.g., food, shelter and clothing but, in line with Kalecki's formulation (see FitzGerald 1990, p. 187, FN: 3), would also include health, education and most manufactured household goods. Moreover, 'essentials' would vary across different cultures and countries and overtime. In the context of a sustainable social policy within a degrowth or zero growth debate, we could argue that production of essential goods needed for achieving social objectives must be maintained whilst resources are directed towards investment goods as well as the non-essential sectors and more importantly towards technologies that would reduce the use of exhaustible resources and greenhouse gas emissions (Berr 2015).

Degrowth or zero growth, however, may well result in unemployment and low utilisation of labour resources especially in countries which are in their early stages of demographic transition and have a growing working age population as is the case in low- and middle-income countries. The high-income countries with large structural unemployment would also have to deal with rising unemployment resulting from a degrowth or zero growth strategy. Some of these results are formalised in a demand-led growth model developed by Fontana and Sawyer (2016) that tracks the paths of growth rates of output, labour resources, capital stock and natural capital depletion. In such an economy in which investment is financed by credit, a high rate of output growth will be matched by a high rate of growth of labour resources (which would lead to a high rate of labour absorption and employment). But this high rate of growth will be at a very high cost in terms of the depletion of natural capital. Lower rates of growth, including zero or negative growth will achieve a low rate of depletion of natural capital but at a low rate of labour absorption and high rate of unemployment (Fontana and Sawyer 2016, Fig. 1, p. 192). The authors correctly argue that the conflict between output growth, on the one hand, and depletion of natural capital and labour absorption, on the other, will not be resolved by market forces and price signals but will require government intervention in regulating credits and directing investment as well as managing the social implications of unemployment.[10] In a similar fashion Taylor, Rezai and Foley (2016) use a demand-driven macroeconomic model linking capital accumulation, economic growth and greenhouse gas accumulation to show that climate change mitigation policies such as a 1% expenditure of world GDP could achieve the greenhouse gas reduction target.

Let us recall that the past and on-going substitution of capital for labour has important implications for the level and composition of

[10] For a radical sociological critic of market-based approach which has the objective of 'resilient communities' for a 'green welfare' see Catney and Doyle (2011).

employment that could be compounded by a degrowth strategy; especially if we note that the impact of degrowth could be cumulative through multiplier effects and impact of growth in the future. It has also been argued that the elasticity of substitution between labour and capital varies with the level of economic development—the lower the level of development or the poorer the country the lower the value of elasticity. It is therefore easier to counter the inequality impacts of degrowth in poorer countries (Jackson and Victor 2016). Be as it may, it does not reduce the need for higher growth in the low- and middle-income countries, a view shared by all sides of the argument of growth versus degrowth.

According to Jackson and Victor (2016), there is potential for maintaining employment in labour-intensive (e.g., health and personal care, education, crafts, catering and culture) sectors in the services sector. Another important policy to reduce the negative impact of low-growth or degrowth on employment is the reduction in the working-week and work-sharing. Zwickl et al. (2016) have argued that 'at a given level of aggregate labour demand, the employment rate might be increased through distributing hours and jobs more equally within the labour force, making work-sharing an important economic policy tool in a low-growth economy' (p. 246). There are also social externalities of health and well-being of workers due to shorter hours and shorter working-week and narrowing down of the gender wag-gap. The argument against work-sharing is focussed on the cost to employers who would be incentivised to replace capital for labour resulting in lower overall employment. But empirical evidence on the negative impact of work-sharing on employment is mixed. In the Great Recession of Germany and the Great Depression of US, work-sharing had a positive impact on employment, whilst in more recent work-sharing schemes of 1980s in the Europe the result has been mixed (ibid.).

In my view and in terms of the potential for employment the health and personal care sectors provide the most potential for increased employment especially if the labour-intensive care of an ageing population is also factored in. A similar argument can be made for a country with a young age structure which could improve the care of its young by devoting more labour to their education and care, particularly for pre-school early education.

Besides the distributional and employment issues, a strategy of low-growth or degrowth must pay attention to the composition of output. Recalling our earlier discussion on the theory of human needs, the composition of output should closely relate to the package of needs and the requirements for their satisfaction at macro-level. Since most 'needs satisfiers' have social and cultural specificity, food and education are just two examples, composition of output would well vary across different countries that depending on local resource availability may well require cooperation, exchange and sharing of resources at international level. Moreover, considering the large energy use and greenhouse impact of intensive agriculture and its commercial distribution there is a need to shift to lower energy use organic agriculture. However, since yield per hectare of organic agriculture currently is lower than chemically driven intensive agriculture, a sustainable agricultural strategy has to plan for increased food production for a growing population but without deforestation and land set aside for preservation. This calls for a sustainable and radical strategy with the objective of moving away from large-scale monoculture of industrial agriculture and towards environmentally sound agricultural production and technology. Such a strategy would focus on biodiversity taking note of local knowledge, synergy and cooperation (Altieri and Nicholls 2005; EU 2019). Moreover, there is also a need to change the composition of agricultural output away from production of feed for the meat industry and towards production of food for human consumption. Such a strategy should also be backed up with promotion of vegetarian diet. Finally, in the discussion of growth, its composition and measurement attention should be paid to not only to the standard index of GDP but also to the complementary indices such as the Index of Sustainable Economic Welfare (Stockhammer, et al. 1997) which takes note of unpaid household labour, social and environmental costs, and income distribution.[11]

[11] For further discussion of the centrality of human needs to a progressive degrowth strategy as well as the need to change composition of output see Koch et al. (2016). See also Wise (2018) for a detailed discussion of the role of biodiverse small-farms in food production and food security around the world, that should be backed by active institutional (technical and financial) support for small farmers as well as paying attention to the role of science and technology in sustainable agriculture to improve productivity and income of farmers and agricultural workers (Baksi 2019).

5 Conclusion

Social policy is by definition societal in the sense that individuals and communities organise activities to achieve certain social objectives, with an implicit or explicit goal of achieving social justice. In the broadest and most positive sense, social policy is about improving human welfare and meeting human needs as well as being concerned with the distribution of resources and assets that would enable individuals to realise their potential, develop their capabilities and contribute to the economic and social life of a country. Social policy thus not only covers the provision of health and education which are about promotion of human capability but also laws, rules and regulations that protect human capability. It is in this context that a sustainable social policy has to be developed that can both address the well-being of people and protect the environment.

At macro-level a sustainable social policy has two important dimensions: growth and time. The growth dimension has to deal with several distributional issues: composition of output and distribution of resources to produce them and distribution of output and resources among different social classes. The time dimension has to deal with the needs of the present and future generations. The two generations have common interests bearing in mind their long temporal and socio-economic overlap and relationships between the two (Stewart 2019; Raworth 2017). The future generation includes the unborn and more importantly the children and youth of today. It should be noted that all generations are heterogenous—socially, economically and culturally and as such may not share the same understanding of environmental problems and sustainability, and have the same interest in sustainable social policies. However, cohorts of generations have worked together for millennia because they share a 'common present' and have needed to create a 'common future'. The economic, social and cultural legacies of the current generation provide a socio-economic foundation for the future generation. There exists an inter-generational contract which is reflected in the culture of 'filial piety' in all societies based on bequests and legacies of the older to the younger generation, household care and mutual support, and formal pension and state/socialised care. Extension of this social contract should be one of the objectives of a sustainable social policy.

Working towards such a society requires a growth strategy that is centred on 'human needs' backed up by a distributional strategy to deliver on people's needs to ensure well-being of current and future generations.

This raises important questions on the composition of output under a low-growth or a degrowth strategy and policies for the distribution of output.

None of the macro-level mitigation social policies of degrowth would be possible without an activist state policy to direct and redirect resources with the engagement and participation of local and citizen-based organisations to ensure the transparency and accountability right through the policy cycle.

The theoretical case for a needs-based growth strategy and a progressive distributional policy is very strong. The organisational, regulatory and financial tax-raising power of state play an important role in delivering on these objectives. We need to build on the struggle of the current generation of the young and not-so-young who have put sustainability on the political agenda and work towards a sustainable social policy agenda in order to create a better sustainable future. We need a paradigm shift in our thinking about economy and society that has been dominated, for far too long, by the economics of consumer sovereignty and individual choice.

References

Ajl, M. (2021) *A People's Green New Deal*, London, London: Pluto Press.

Altieri, M. A. and Nicholls, C. I. (2005) *Agroecology and the Search for a Truly Sustainable Agriculture*, Mexico: Environmental Training Network for Latin America and the Caribbean, United Nations Environment Programme. http://www.agroeco.org/doc/agroecology-engl-PNUMA.pdf. Accessed: 21 June 2021.

Angus, I., Kovel, J., Hartmann, B. and Butler, S. (2011) *Population, Immigration and Environmental Crisis*, Chicago: Haymarket Books.

Baksi, S. (2019) 'Sustainable Agriculture and Family Farmers,' *Review of Agrarian Studies*, vol. 9, no. 2, July–December, pp. 114–120 [Review of T. A. Wise (2018) Eating Tomorrow: Agribusiness, Family Farmers, and the Battle for the Future of Food, The New Press, New York].

Berr, E. (2015) 'Sustainable Development in a Post Keynesian Perspective: Why Eco-development Is Relevant to Post Keynesian Economics,' *Journal of Post-Keynesian Economics*, Vol. 37, pp. 459–480.

Briggs, A. (1961) 'The Welfare State in Historical Perspective,' *European Journal of Sociology*. Abridged and printed in Pierson, C., Castles, F. G. and Naumann, I. (eds.) (2014) *The Welfare State*, 3rd edition, Cambridge: Polity Press, pp. 14–27.

Buchs, M., Bardsley, N. and Duwe, S. (2011) 'Who bears the brunt: Distributional effects of climate change mitigation policies.' *Critical Social Policy*, 31(2), 285–307.

Catney, P. and Doyle, T. (2011) 'The Welfare of Now and the Green (Post) Politics of the Future,' *Critical Social Policy*, Vol. 31, No. 2, pp. 173–192.

Dinda, S. (2004) 'Environmental Kuznets Curve Hypothesis: A Survey,' *Ecological Economics*, Vol. 49, pp. 431–455.

Eckersley, R. (2004) *The Green State: Rethinking Democracy and Sovereignty*, Cambridge, MA, USA: The MIT Press. https://doi-org.eur.idm.oclc.org/10.7551/mitpress/3364.001.0001.

Ehrlich, P. R. and Ehrlich, A. H. (1968) *The Population Bomb*, New York, US: Ballantine Books.

Ehrlich, P. R. and Ehrlich, A. H. (1970) *Population, Resources, Environment: Issues in Human Ecology*, San Francisco, CA: Freeman.

Environment and Development Economics (1998) Special Issue on Environmental Kuznets Curve, 2 (4).

Esmailian, S. (2015) 'Out of Frying Pan: People on the Move in the Warmer World and How to Receive,' in K. Borgnas, T. Eskelinen, J. Perkio and R. Warlenius (eds.) *The Politics of Ecosocialism: Transforming Welfare*. Routledge, pp. 148–162.

Esping-Andersen, G. (1990) *The Three Worlds of Welfare Capitalism*. Cambridge: Polity Press.

EU (2019) *Position Paper on Agroecology. Organic and Agroecology: Working to Transform Our Food System*. https://www.organicseurope.bio/content/uploads/2020/06/ifoameu_position_paper_agroecology.pdf?dd. Accessed: 21 June 2021.

FAO (2021) *Food Outlook—Biannual Report on Global Food Markets*. Rome, Italy: FAO. http://www.fao.org/documents/card/en/c/cb4479en/. Accessed: 26 June 2021.

FAO (2002) World Agriculture: Towards 2015/2030, Summary Report. Rome, Italy: FAO http://www.fao.org/3/y3557e/y3557e00.pdf. Accessed: 30 March 2021.

FAO (2009) *Global Agriculture Towards 2050*. Rome, Italy: FAO. http://www.fao.org/fileadmin/user_upload/lon/HLEF2050_Global_Agriculture.pdf. Accessed: 30 March 2021.

FitzGerald, E. V. K. (1990) 'Kalecki on Financing Development: An Approach to the Macroeconomics of the Semi-industrialised Economy,' *Cambridge Journal of Economics*, June, Vol. 14, No. 2 (June 1990), pp. 183–203.

Fitzpatrick, T. (1998) 'The Implications of Ecological Thought for Social Welfare,' *Critical Social Policy*, Vol. 54, No. 18 (1), pp. 5–26.

Fitzpatrick, T. (2011) *Understanding the Environment and Social Policy*, Bristol, UK: Policy Press.

Fontana, G. and Sawyer, M. (2016) 'Towards Post-Keynesian Ecological Macroeconomics,' *Ecological Economics* 121, pp. 186–195. https://doi.org/10.1016/j.ecolecon.2015.03.017.

Foster, J. B. (1998) 'Malthus' Essay on Population at Age 200: A Marxian View', *Monthly Review*, Vol. 50, No. 7. Dec.

Fraser, N. (1989) *Unruly Practices: Power, Discourse and Gender in Contemporary Social Theory*. Minneapolis, US: University of Minnesota Press.

Ghosh, J. (2011) 'Michael Kalecki and the Economics of Development,' *Monthly Review*. Accessed online: http://mrzine.monthlyreview.org/2011/ghosh270511p.html.

Gough, I. (2015) 'Climate Change and Sustainable Welfare: The Centrality of Human Needs,' *Cambridge Journal of Economics*, Vol. 39, pp. 1191–1214.

Hardin, G. (1968) 'The Tragedy of the Commons.' *Science* 13 Dec.: Vol. 162, Issue 3859, pp. 1243–1248. https://doi.org/10.1126/science.162.3859.1243

Hickel, J. (2019) 'The Contradiction of the Sustainable Development Goals: Growth Versus Ecology on a Finite Planet,' *Sustainable Development*, pp. 1–12.

Hickel, J. (2020) *Less Is More*, London: William Heinemann Publishers.

IPCC (2018) Global Warming of 1.5°C. An IPCC Special Report on the impacts of global warming of 1.5°C above pre-industrial levels and related global greenhouse gas emission pathways, in the context of strengthening the global response to the threat of climate change, sustainable development, and efforts to eradicate poverty. https://www.ipcc.ch/sr15/chapter/spm/. Accessed: 21 June 2021.

Jackson, T. and Victor, P. A. (2016) 'Does Slow Growth Lead to Rising Inequality? Some Reflections and Numerical Simulations,' *Ecological Economics* 121, pp. 206–219.

Jawad, R., Jones, J. and Messkoub, M (eds.) (2019) *Social Policy in the Middle East and North Africa: The New Social Protection Paradigm and Universal Coverage*, Cheltenham, UK: Edward Elgar Publishing Ltd. I–IX, 1–255 pp.

Kalecki, M. (1976) *Essays on Developing Economies*, Brighton, UK: Harvester.

Koch, M., Buch-Hansen, H. and Fritz, M. (2016) 'Shifting Priorities in Degrowth Research: An Argument for the Centrality of Human Needs,' *Ecological Economics*, Vol. 121, pp. 74–81.

Kwon, H.-j., Mkandawire, T. and Palme, J. (2009) 'Introduction: Social Policy and Economic Development in Late Industrializers.' *International Journal of Social Welfare*, 18: S1–S11.

Mkandawire, T. (ed.) (2004) *Social Policy in a Development Context*. Geneva: UNRISD.

Malthus, T. R. (1798) *An Essay on the Principle of Population and Other Writings*. (Republished in Penguin Classics, Paperback—2015)

Marshall, T. H., ed. (1950) *Citizenship, and Social Class and Other Essays.* Cambridge, CUP. Reprinted in Pierson, C., Castles, F. G. and Naumann, I. (eds.) (2014) *The Welfare State Reader*, 3rd edition, Cambridge: Polity Press, pp. 28–37.

Mehri, N., Messkoub, M. and Kulke, S. (2020) 'Trends, Determinants and The Implications of Population Aging in Iran.' With N. Mehri and *Ageing International*, No. 45, pp. 327–343. https://doi.org/10.1007/s12126-020-09364-z.

Messkoub, M. (1999) 'Crisis of Ageing in Less Developed Countries: Too Much Consumption or Too little Production.' *Development and Change*, Vol. 30, No. 2, April.

Messkoub, M. (2021) 'Social Policy in the MENA Region,' in H. Hakimian, ed. (2021) *Routledge Handbook on Middle Eastern Economy*, July, London: Routledge.

Nordhaus, W. (2008) *A Question of Balance: Weighing the Options on Global Warming Policies,* New Haven, CT, US: Yale University Press.

Patel, R. and Moore, J. (2017) *A History of the World in Seven Cheap Things: A Guide to Capitalism, Nature, and the Future of the Planet*, Oakland, CA: University of California Press.

Pearce, D., Markandya, A. and Barbier, E. B. (1989) *Blue Print for a Green Economy: A Report*, London: Earthscan.

Pearce, D. (1992) 'Green Economics,' *Environmental Values*, Vol. 1, no. 1, pp. 3–13. http://www.environmentandsociety.org/node/5454.

Pierson, C. (1998) *Beyond the Welfare State*, 2nd edition, Cambridge: Polity Press.

Pierson, C., Castles, F. G. and Naumann, I. (eds.) (2014) *The Welfare State*, 3rd edition, Cambridge: Polity Press.

Piketty, T. (2014) *Capital in the Twentieth Century*, Cambridge, MA, US: Harvard UP.

Raworth, K. (2017) *Doughnut Economics: Seven Ways to Think Like a 21st-Century Economist*, London: Random House Business.

Redclift, M. (2018) 'Sustainable Development in the Age of Contradictions,' *Development and Change*, Vol. 49, No. 3, pp. 695–707.

Rezai, A. and Stagl, S. (2016) 'Ecological Macroeconomics: Introduction and Review,' *Ecological Economics*, Vol. 121, pp. 181–185.

Robra, B. and Heikkurinen, P. (2019) 'Degrowth and the Sustainable Development Goals,' in W. Leal Filho et al. (eds.) (2019) *Decent Work and Economic Growth*, https://doi.org/10.1007/978-3-319-71058-7_37-1.

Sawyer, Malcolm C. (1989) *The Challenge of Radical Political Economy : An Introduction to the Alternatives to Neo-classical Economics.* Hemel Hempstead, England: Harvester Wheatsheaf.

Stern, N. (2009) *A Blueprint for a Safer Planet*, London: Bodley Head.

Stewart, F. (1985) *Basic Needs in Developing Countries*, Baltimore, US: Johns Hopkins University Press.

Stewart, F. (2019) 'The Human Development Approach: An Overview,' *Oxford Development Studies*, Vol. 47, No. 2, pp. 135–153.

Stockhammer, E., Hochreiter, H., Obermayr, B. and Steiner, K. (1997) 'The Index of Sustainable Economic Welfare (ISEW) as an Alternative to GDP in Measuring Economic Welfare. The Results of the Austrian (revised) ISEW Calculation 1955–1992,' *Ecological Economics*, Vol. 21, No. 1, April, pp. 19–34.

Taylor, L., Rezai, A. and Foley, D. A. (2016) 'An Integrated Approach to Climate Change, Income Distribution, Employment, and Economic Growth,' *Ecological Economics*, Vol. 121, pp. 196–205.

Thirlwall, A. P. and Pacheco-Lopez, P. (2017) *Economics of Development: Theory and Evidence*, 10th edition, Basingstoke, UK: Macmillan.

Weber, H. and Dabbs Sciubba, J. (2019) 'The Effect of Population Growth on the Environment: Evidence from European Regions,' *European Journal of Population*, Vol. 35, No. 2, pp. 379–402. https://doi.org/10.1007/s10680-018-9486-0.

Wiggins, D. (2005) An Idea We Cannot Do Without: What Difference Will It Make (eg. to moral, political and environmental philosophy) to Recognize and put to Use a Substantial Conception of Need? *Royal Institute of Philosophy Supplements*, Vol. 57, pp. 25–50. https://doi.org/10.1017/S1358246100009139.

Wise, T. A. (2018) *Eating Tomorrow: Agribusiness, Family Farmers, and the Battle for the Future of Food*, The New Press, New York.

WCED—World Commission on Environment and Development [Bruntland Report] (1987) Our Common Future. Geneva, Switzerland, 400 pp. https://doi.org/10.2307/2621529.

Zwickl, K., Disslbacher, F. and Stagl, S. (2016) Work-Sharing for a Sustainable Economy *Ecological Economics* 121, pp. 246–253. https://doi.org/10.1016/j.ecolecon.2015.06.009.

Moving Towards a Sustainable Economy—A Social Liberal View

Heinrich Bortis

1 Introduction: Problem and Plan

To move in the direction of a sustainable economy and attempting to fully realise sustainability requires a new socio-economic paradigm, *social liberalism* to wit, that is, a fully fledged alternative to the doctrines of *liberalism* and *socialism*, historically realised through capitalism, now dominated by huge corporations and big finance, and socialism with central planning.

To be sure, the various strands of the green movement undertake genuine efforts to harmonise the relationship between man (society) and nature. And within the capitalist system, efforts are made to prevent a climate catastrophe in order to be able to maintain the system. However, there are also strong adverse forces at work. The struggle for survival and

I am indebted to my former scientific collaborator Andreas von Ah for pertinent remarks on controversial issues. Of course, all responsibility remains mine.

H. Bortis (✉)
University of Fribourg, Fribourg, Switzerland
e-mail: heinrich.bortis@unifr.ch

N. Karagiannis and J. E. King (eds.), *Visions and Strategies for a Sustainable Economy*, Global Institute for Sustainable Prosperity,
https://doi.org/10.1007/978-3-031-06493-7_9

the striving for high profits leads on to using fossil energy to be able to produce as cheaply as possible; driven by the profit motive, multinationals and transnationals are looking for cheap labour, which, in many instances, is put to work in industry and mining under difficult conditions, with the natural environment being frequently heavily damaged through pollution and production of waste. However, given the global imbalances and widespread poverty, the majority of people simply cannot afford to rely on alternative energy sources to fossil fuels. In agriculture, the introduction of high yielding varieties drives up the use of water and of pesticides, favouring large scale farms and tending to eliminate family peasants; moreover, animals are raised in conditions of mass production in an entirely unnatural way and plants genetically modified, with largely unknown (long term) consequences for the health of human beings. Moreover, deforestation creates the conditions for industrial and agricultural mass production, accompanied by an almost boundless growth of cities. Given all this, global warming goes on, biological diversity is reduced and many species are threatened by extinction.

In this chapter, we summarise all these environmental problems under the heading of *environmental alienation*. This notion sums up the overall deviation from an ideal sustainable economy, which implies harmony between man (society) and nature.

However, this chapter does not deal with technical aspects of sustainability. Instead, the paper focuses on an additional impetus to be given at moving towards a sustainable economy through implementing, step by step, a new paradigm, social liberalism, which is, in our view, badly needed to permanently bring about a sustainable economy.

In the second section, we very briefly deal with the social philosophies of liberalism and social liberalism. This is followed by opposing neoclassical economics, the economic theory of liberalism, to classical-Keynesian political economy, representing the economic theory of social liberalism. Fourth, the notions of the natural state and alienation are hinted at. In the fifth section, the relationship between classical-Keynesian political economy and the natural environment is alluded to. Sixth, the actually prevailing environmental situation is sketched on the basis of the external employment mechanism. In the seventh section, normative aspects of sustainability are taken up, grounded upon the internal employment mechanism. Section eight deals with environmental policies to be pursued. The conclusion (section nine) is about mobilising the social capacities of human beings to create the enthusiasm required to speed up and to intensify the movement towards a sustainable economy.

2 Liberalism and Social Liberalism

The autonomous and self-contained individual stands at the centre of the social philosophy of Liberalism. Social phenomena are derived and come into being through explicit and formal or implicit and informal contracts, firms and loose associations would be instances. Since the individuals are primary, institutions in the liberal sense could be called individualistic institutions. The most important, almost natural, liberal institution is the market, which, ideally, co-ordinates the rational, profit and utility maximising behaviour of producers and consumers, respectively, in a socially optimal way. In a liberal society, historically realised through Capitalism, individualistic materialism dominates and there is a tendency for everything to become a commodity. The state should, ideally, be of a minimum size, its main task being to ensure competition and to provide for some public goods, for instance, education, a legal system and internal and external security.

With Social Liberalism individuals are *social* individuals. This implies that there are "social states of affairs", the "social" for short, resulting in the existence of genuinely social entities, that is, *social institutions*. Such institutions come into being if several or all members of a society persistently pursue *common or social aims* that isolated individuals could not achieve. In doing so, individuals or groups exercise differing complementary functions (planning or executive, physical or intellectual) within a social institution; co-operation and co-ordination are essential if such institutions are to function properly. In this sense, football teams, orchestras and enterprises are social institutions. But the most striking example of a social institution is the social process of production: in a monetary production economy based upon extensive division of labour, production of commodities goes on by means of commodities and labour; each sector of production and each enterprise (themselves social institutions), and each individual performs a specific function within the process of production, and thus contributes to reaching a common (social) aim, that is the production of the social surplus and of the social or national product.

Besides the social institutions there are behavourial institutions, customs and habits in the domains of consumption, culture and ethics. However, behaviour takes place on the basis of specific ideas and values in the economic, social, political, cultural and religious domain. These interdependent ideas and values make up the intellectual scenery of some time period.

Social and behavioural institutions are all complementary, making up society and the state and bringing about a certain way of life. Given this, society and the state are much more than the sum of the individuals living on the state territory. Consequently, societies and states conceived along social liberal lines are unique. This, in turn, is associated with cultural diversity and the possibility of mutual enrichment in the cultural and material domains.

3 Neoclassical Economics and Classical-Keynesian Political Economy

With liberal economic theory, neoclassical economics to wit, all the great economic problems—value, distribution and employment most importantly—are, ideally, solved on competitive markets. The rational, profit and utility maximising behaviour of the economic agents are supposed to be co-ordinated by interdependent markets in such a way that a social optimum obtains, the Pareto optimum: the situation of no agent can be improved without worsening the situation of some other agent. The Pareto optimum implies a general equilibrium in all markets, hence full employment of all factors of production, land, capital, and, most importantly, labour. The principle of supply and demand closely associated with the marginal principle underlies the neoclassical equilibrium model.

The economic theory of social liberalism is classical-Keynesian political economy (Bortis 1997). The economy considered is a *monetary production economy* in which the social process of production is at the centre of the economic system; production has to be financed, hence the crucial importance of the financial sector and production has to be sold against money, hence the overriding importance of demand in money terms.

Given this, the basic classical-Keynesian model must be made up of the primary principles governing the functioning of modern monetary production economies. Two principles are of classical origin: the labour value principle summarises the essential features of the immensely complex social process of production to provide the essence of the prices of production, which are the fundamental prices in a monetary production economy; the surplus principle of distribution implies that the distribution of income is positively a problem of social power and normatively of distributive justice situated at the heart of social ethics. Keynes provided a third principle, namely the principle of effective demand, related to determining the scale of economic activity. These three principles imply that money plays a fundamental role. Indeed, the processes of production and circulation could not go on without money, since production takes time, outlays and receipts are not synchronised and goods are never exchanged against other goods but always against money, which also acts as a store of value and as such is intimately connected to the financial sector.

The classical-Keynesian core model consists of a price and of a quantity equation (Bortis 2019, pp. 194–96). The price equation embodies two principles of classical political economy: the labour value principle and the surplus principle. The quantity equation, the super-multiplier, tells us how output and employment are determined in principle through effective demand. Private and public consumption represents the internal demand components of demand, whereby private consumption increases if income distribution becomes more equal. Exports obviously lead on to an increase of external demand, which is reduced by imports.

4 The Natural State and Alienation

The notion of alienation has been coined by Karl Marx—through reinterpreting Hegel—to denote cleavages between natural states of societies and social reality. In his *Frühschriften* Marx equates the natural state with communism; in fact, he puts communism equal to humanism and naturalism. The old Liberals, Adam Smith, in particular, considered the (ethically grounded) market economy as a natural state of affairs; however, Adam Smith's ethical foundations of the market gradually faded away to be replaced by the notion of perfect competition; here the desirable state is represented by the Pareto optimum, which is entirely individualistic and devoid of ethical content.

The social liberal notion of the *natural state* is grounded upon the fundamental social ethical value of the *Common Good*, which is "social" in the sense proper, that is, in the social liberal sense as is sketched in Sect. 2 above. Common aims are to be reached through common complementary actions which require co-operation and co-ordination. Common aims are pursued, for instance, within firms: the production of one or several goods, or within the social process of production, resulting in the social product and the social surplus.

In the sphere of the material (economic) basis, the most important components of the Common Good are full employment or the absence of involuntary unemployment, and a fair distribution of incomes and wealth. The institutional social, political, legal, ethical, cultural and religious superstructure ought to be such that the space of freedom is maximised and that the social individuals may prosper, that is, develop their dispositions and acquire new capacities. Given this, the political economy of social liberalism, classical-Keynesian political economy to wit, could also be called the political economy of the Common Good, which is, in fact, the social liberal equivalent to the liberal Pareto optimum.

In capitalism in general, the immensely complex phenomenon of alienation appears, for example, in the form of involuntary unemployment, precarious living conditions due to low wages and inappropriate work conditions. In the capitalist industry, an important dimension of alienation shows up in the workers becoming appendages of the machine, beautifully illustrated by Charlie Chaplin's *Modern Times*.

All the elements of alienation mentioned in the introduction to this chapter represent environmental alienation, defined as the deviation from the sustainable (natural) state, which implies harmony between man (society) and nature. In general, the huge environmental problems, global warming in the first place, are due to large scale alienation between man (society) and nature brought about by capitalism in general and by neoliberal capitalism in particular. In the following sections, we now attempt to come to grips with environmental alienation in capitalist economies through putting to use classical-Keynesian tools of economic analysis.

5 Classical-Keynesian Political Economy and the Problem of the Natural Environment

To apply classical-Keynesian political economy to the problems of the natural environment and to a sustainable economy, two output and employment mechanisms contained in the super-multiplier equation, the classical-Keynesian quantity equation, have to be considered (Bortis 1997, pp. 190–98; 2019, p. 197):

The *internal* mechanism is based upon the fundamental macroeconomic equality ($S = I$):

$$Q_i = \frac{G}{(1 - c) - a} \tag{1}$$

The economy is set into motion through government expenditures (G). The level of output and employment (Q_i) is governed by (G) and the consumption/output ratio (c), which, in turn, depends on the distribution of incomes. In a Keynes/Beveridge vein, a more equal income distribution enhances the spending power of the population, raising thus the output and employment level. The gross investment volume is, like consumption, a derived variable; hence, the gross investment/output ratio (a) is given.

The *external* mechanism is based upon the current account equilibrium ($X = M = b\,Q$):

$$Q_e = (1/b)X \tag{2}$$

The inverse of the import/output ratio (b) represents the familiar export multiplier stating how output (Q_e) and thus employment are governed by the external employment mechanism.

If Q_e exceeds Q_i, a surplus in the balance of current account occurs, and vice versa. This emerges from the super-mulitiplier equation (9.2a) set out in Bortis (2019, pp. 198 and 205). With the balance of current account in equilibrium, the super-multiplier relation equals relation (1) above, picturing the internal employment mechanism.

The internal employment mechanism, which might be called the Common Good mechanism, is the starting point to consider the natural state, implying harmony between man (society) and nature, which is of a normative and desirable nature. The external mechanism, however, enables us to deal with the actually prevailing situation of environmental alienation, picturing a heavy disturbance of the relationship between man (society) and nature. It is to this situation we now turn.

6 The External Employment Mechanism at the Origin of Environmental Problems

In a globalised world, it is exceedingly difficult to pursue economic policies based upon the internal output and employment mechanism (relation (1) above). Indeed, to enhance effective demand through private and public consumption requires distribution policies, possibly resulting in higher wages and, eventually, higher taxation, reducing thus international competiveness; moreover, increasing effective demand may lead to deficits on current account. Given this, all countries, more or less integrated into the globalised world economy, will rely on the external employment mechanism to enhance output and employment (relation (2) above).

According to the external employment mechanism each country attempts to create workplaces through exporting as much as is possible. This mechanism is greatly enhanced through globalisation and large free-trade areas, like the European Union. A world currency based upon a national currency, the US dollar to wit, also strengthens the external employment mechanism, with the deficit countries facing ever-higher indebtedness.

Given the fact that, in capitalist economies, there is no tendency towards a full-employment equilibrium whatsoever, competition will not only operate through a favourable price/quality relationship, but will consist in a struggle for market shares and, in the extreme, competition will result in a merciless struggle for survival. And the basic aim of Capitalism, above all of Monopoly-Finance Capitalism, is large profits in the real sector, and profit-making is enhanced by the activities of the financial sector, by means of takeovers for example.

Given this, the working mechanism of the globalised capitalist system leads to massive alienation in the socio-economic sphere (large involuntary unemployment, precariousness, growing inequalities in income and wealth distribution). However, massive alienation also occurs in the environmental domain and, therefore, deeply disturbs the relationship between man (society) and nature; in fact, intensive competition and struggles for market shares and survival on the world markets inevitably lead on to considering nature as a free good.

Moreover, the structure of foreign trade leads on to the formation of a small group of highly industrialised aristocratic countries—Germany, Japan and Switzerland as the leading examples—and to a large number of proletarian countries where highly industrialised isles may come into being in some instances. "[Indeed] the employment effect of foreign trade will be particularly strong if the bulk of exports consists of high-quality industrial products and services and if imports are, in the main, made up of primary goods; in this case the terms of trade will, as a rule, be favourable. High-quality industrial goods and services are, as Nicholas Kaldor has emphasised time and again, labour-intensive if account is taken of direct, indirect and past labour (fixed capital) and of research and development activities. However, primary products (agricultural products, raw materials and energy resources) are land-intensive and industrial standard products create few and low quality workplaces since their production is usually very mechanised with machines being imported as a rule. Here the employment effect of foreign trade will be weak and heavy outside dependence regarding technologically advanced industrial products will occur" (Bortis 2019, p. 199).

Moreover, socio-economic alienation is, on the one hand, strongly reinforced by the presence of massive involuntary unemployment worldwide which results in a pressure on wages. On the other hand, there is rapid technical progress which benefits all large firms. These are able to set relatively low prices and simultaneously achieve large profits. This drives part of small and medium-sized enterprises out of the market, weakening thus the position of the middle classes. These developments are enhanced through the big players in the financial sector attempting to increase profits through takeovers. Ultimately, *persistent* heavy involuntary unemployment inevitably leads to ever-growing inequalities in the distribution of incomes and wealth *worldwide*.

Furthermore, alienation steadily increases through a contradiction between the internal and the external development mechanism on the world level (Bortis 2019, pp. 199–201): Each region and each country attempts to attract enterprises who export the bulk of their production. Given this, taxes must not be too high relative to other regions and countries. This may lead to stagnating and even declining government expenditures worldwide. Since, however, world economic activity must be governed by the internal output and employment mechanism, world effective demand is reduced which renders competition and the associated struggle for survival even more ferocious. Finally, through the external employment mechanism the successful exporters of high-quality industrial goods and services may nevertheless enjoy a satisfactory, even a booming economic situation. The losers, however, mostly countries exporting agricultural products, raw materials and energy resources, and, eventually, standard industrial products, will be precipitated into the abyss of mass unemployment, increasing indebtedness and social and political instability. Owing to the law of increasing returns and huge differences in technological standards and dynamism as well as to the principle of effective demand, there is a *cumulative causation of growing disequilibria and widening inequalities*, resulting in ever-larger inequalities of income and wealth, and employment opportunities worldwide, an argument put forward by Friedrich List, Gunnar Myrdal and Nicholas Kaldor (Bortis 2019, p. 200); regarding the inequality of incomes and wealth wide-ranging empirical work has been done, James K. Galbraith and

Thomas Piketty being important authors. The inequalities in the distribution of incomes and wealth as well as the existence of massive involuntary unemployment strongly enhance overall alienation, socio-economic and environmental.

Alienation can only be systematically tackled if we know what the natural or normative counterpart to alienated reality looks like. To this issue we now turn.

7 The Internal Employment Mechanism, the Sustainable Economy and the Natural State

To realise the sustainable economy requires the implementation of the internal output and employment mechanism inherent in classical-Keynesian political economy (relation (1) above and Bortis 2019, pp. 201–4). This has far-reaching implications. Indeed, to apply the internal employment mechanism requires moving from the actually prevailing neoliberal Capitalism to Social Liberalism. This means fundamental changes in the value system of modern societies amounting to a *Second Great Transformation*, following up Karl Polanyi's *First Great Transformation* 1750–1830 which brought about the transformation from a traditional society to a modern industrial-cum-service society. In modern capitalist societies the economy (the market) and materialistic values—consumerism and striving for profits and wealth—are primary and cultural–ethical values are secondary. *In social liberalism this order of values is reversed:* the economy, the material basis, becomes ancillary and the use of the social surplus ought to be directed towards cultural and ethical purposes through the institutional superstructure, implying that the social and political sciences, political economy most importantly, become moral sciences; in fact, the striving after perfection in all domains and to realise the Common Good to the largest possible extent would move to the fore. This was, broadly, also Keynes' view who has perceived with incomparable clarity that the *materialist capitalist era* must be followed by an *epoch dominated by ethics and culture* if modern civilisation is to survive. On this, the Italian Keynes biographer Piero Mini writes: "[Even the] most superficial reading of Keynes's writings [...] should convince anybody that Keynes was not an economist as we understand the term. He was primarily a social philosopher, a cultural leader interested in the cultural amelioration of society. Throughout his life he prodded the people and their leaders to set for themselves standards worthy of men [...]: the promotion of solidarity among people (the opposite of Benthamite individualism and egoism) and the extension of the realm of beauty (the opposite of Benthamite "push-pin"). Attainment of full employment—via the agency of the state and through substantial reforms of the system—was to be the way of attaining these [...] ends" (Mini 1991, pp. 102/3). "[Keynes was greatly influenced] by an anti-rationalistic current associated with certain critics of the emerging commercial England [e.g. Coleridge and Carlyle, who]" (Mini 1991, p. xvii) "stressed the primacy of the spiritual over the material, of ends over means, of intuition over the narrowly logical. They were humanists who opposed the claims of [materialistic] individualism with the claims of community and tradition and who had a positive view of the state and of the binding value of culture" (Mini 1991, p. 2).

This reversal of values implies that the education system should be of a traditional nature, with an emphasis on culture (*Bildung*): history of facts and ideas, ancient and modern philosophy and the fine arts, to give instances. *Bildung* will have a positive influence on professional formation in all domains (*Ausbildung*), as an engineer or an economist, for example, in that autonomous and creative thinking is greatly enhanced.

As suggested above, the fundamental and all-encompassing social ethical value of social liberalism is the Common Good. The natural state is an objective state of affairs in which the Common Good is realised as far as this is possible for imperfect human beings. Alienation, socio-economic and environmental, briefly dealt with in the preceding section, represents a deviation from the natural state.

The Common Good comprises values related to the material basis of a society and to the socio-political and cultural–ethical superstructure. The most important values linked with the material basis are full employment, the absence of involuntary unemployment, a fair distribution of incomes and socially appropriate prices. According to the internal employment mechanism, full employment can be reached through bringing about sufficient effective demand created by government expenditures and private consumption, which, in turn, depends on an equitable distribution of incomes, based upon the principle of distributive justice related to relative wages and a socially appropriate profit rate; with distribution regulated, socially appropriate prices of production will obtain (Bortis 2019, pp. 201–4).

Let us now consider some crucial issues related to a sustainable economy taken up in the present volume.

First, a sustainable economy may require a stationary economy and, in a post-Keynesian vein, this would render it impossible to realise profits since the rate of profits is governed by the rate of growth. However, according to social liberal classical-Keynesian political economy where the share of property income is regulated by the surplus principle, any rate of profits is possible even if there is no growth. Indeed, in the zero growth case, profits and the surplus in general, would simply be consumed, spent on cultural purposes and paid on taxes. Saving would have to be zero in order to satisfy the macroeconomic equilibrium condition "saving equal investment" (see on this equation (7b) and its implication in Bortis 1997, pp. 168–69). In any case, in a social liberal economy, the rate of growth compatible with sustainability may be chosen.

Second, full employment is an essential component of the Common Good. This enables the state to *minimise* social and welfare expenditures. In fact, people able to work do not want to be dependent upon unemployment benefits; they simply want to work and to get a fair salary. Given this, the bulk of social expenditures will consist of expenditures in the health sector.

Third, in a Common Good perspective, the economy and technology must be in the service of Man and Society. This implies that modernity must be based upon the achievements of tradition, and these achievements have to be preserved. Very importantly, artisans and family farmers have accumulated an immense knowledge over very long periods of time, which has to be maintained and perpetuated. This knowledge also contains precious information about the natural environment and how to bring about a sustainable economy. Hence craftsmen and family farmers could greatly contribute to building up a harmonious relationship between Man (Society) and Nature. In fact, craftsmanship and traditional farming are basic human activities constituting autonomous fundamental human values, indispensable for the prospering of the social individuals. Modern methods of production, machinery and associated to artificial intelligence are derived activities. If the basis is damaged and even destroyed, then, in the long term, the modern superstructure will break down, causing heavy damage to man, society and nature. Consequently, socio-economic and environmental alienation will grow dramatically.

Fourth, the importance of tradition also implies a certain attitude towards modern technologies, artificial intelligence (algorithms, robots) most importantly. Indeed, social individuals get more perfect through social activities, for example, going to school, to university, discussing, reading, contemplating works of art and architecture, practising sports, and, last, but not least, the enhancing of manual skills; in a world with ever scarcer natural resources, skilled trades—craftsmanship—might become of crucial importance again; Richard Sennett's very important book *The Craftsman—Handwerk* (Sennett 2008) points in this direction. In this context, we may mention that, in a social liberal society in which cultural–ethical values are primary, craftsmen will be badly needed to maintain the immense cultural heritage (buildings, paintings and other) to be found in Europe and elsewhere.

Sennett's book leads to a very important point. At present traditional types of manual work, traditional craftsmanship and traditional farming in the main, but also service activities, are threatened by the invasion of computer software (artificial intelligence and the use of robots, for example). If going on in an unrestricted way these developments are bound to be destructive for modern societies, since they destroy precious *basic* knowledge (*savoir-faire,* know-how) acquired by artisans and traditional farmers over hundreds and thousands of years.

On this issue, the former Fribourg (Switzerland) Professor of Statistics and Operations Research, *Ernst Billeter,* founder of the *first* institute of computer sciences in the world in 1958, told his students around 1965 something like this: "Computer sciences, particularly software applications, must be the object of *post-graduate* studies in all domains. First, our young people must go through a *traditional* apprenticeship or obtain a *traditional* university degree with an emphasis on *Bildung* (culture) leading on to autonomous and comprehensive thinking (*eigenständiges und ganzheitliches Denken*), and only subsequently they ought to become acquainted with computer sciences. This will enable artisans, traditional farmers and holders of a university degree to use computer software in a socially appropriate, that is, a Common Good oriented way."

In our view, Billeter's statement is at present, at the beginning of the twenty-first century, more valid than ever. However, it must be adapted to actually prevailing circumstances. The computer sciences have by now become a fully-fledged scientific system. Given this, post-graduate courses in computer sciences for artisans, political economists, social and political scientists as well as natural and medical scientists ought to provide a general introduction to the computer sciences. This introduction should enable artisans and general academics to discuss with computer scientists about the introduction of software technologies, artificial intelligence most importantly, in the economic, social and political domain as well as in the field of natural, technical and medical sciences. The question to be answered will always be the same: does a specific software application *promote the Common Good* or not? The answer to this question also implies that *the destruction of "human agency"* is to be avoided and that *the prospering of the social individuals is to be promoted. The decision to introduce a specific software application should always be taken by traditional artisans and academics.*

The absolute necessity to adopt this procedure is confirmed by the growing number of very critical publications on *artificial intelligence (AI)*, Daneke (2020) being an excellent example. To be sure, there are positive aspects, if AI is used to improve *techniques*, enhancing thus the capacity of man: "[For instance,] powerful medical diagnostics (especially using radiological and biopsy data) are already well established and new treatments are also emerging on a regular basis" (Daneke, p. 23). However, when complex social, macroeconomic, political and cultural problems are dealt with, precise scientific methods turn out to be inadequate. Indeed, based on holistic thinking (*ganzheitliches Denken*), *understanding* is required which is of an essentially qualitative nature, metaphysical elements, for example, the vision of man and society enter the analysis; moreover, a search for the most plausible principles of analysis and policy action is required in the social and political sciences and political economy, most importantly; given all this, the conclusions, while being rational, will, as a rule, not be certain, implying that knowledge is probable in a qualitative sense as has been emphasised by Maynard Keynes (on this see O'Donnell (1989), specifically Chapter 2, pp. 28–49). All this defies the systematic use of artificial intelligence and gives way to *common sense,* put to the fore by Keynes and Einstein. Indeed, "[the] divergence from the isomorph with human capabilities will be pretty much moot as machine intelligence speeds away. ... Along the way, however, practitioners of machine learning might lose track of some of the vital understandings and enhancements that they originally promised. If they have already lost interest in arriving at a full understanding of the human mind and the meaning of consciousness, they might also inadvertently obliterate many ethical and legal considerations in their haste to monetise 'superintelligence'. ... [Surveys among AI pioneers, business leaders and policy-makers have] identified a number of negative prospects [of AI]. In addition to privacy and imbedded algorithmic biases, *the destruction of 'human agency' was paramount* [our emphasis]. Given macroeconomic mismanagement and political disintegration, individuals in developed nations have never felt such a sense of loss of control over their lives, and most experts agree that widespread use of AI will make matters much worse. Toxic levels of anomie and alienation could have immense societal consequences. While the experts have diagnosed the loss of norms [and] personal efficacy ... [a lack] of understanding of the social ecology of AI tends to play down its central, yet sublime theme. On its current institutional trajectory, AI is not so about enhancing human intelligence as it is about insisting that humans become more like machines themselves, and reorder their lives to become more compliant cogs in the bigger machine. Once algorithms know us better than we know ourselves, many an on-going effort to interject legal and ethical concerns might fall by the wayside" (Daneke 2020, pp. 23–24).

It is likely, that the unrestricted application of AI on exclusively money-making lines will lead to reduce or even to ignore environmental problems since the productive use of nature will be considered costless and the notion of *external effects* might definitely become a foreign word. Moreover, it is highly likely that, given an extended use of AI, involuntary and structural unemployment will dramatically increase. Given this, the actually prevailing external development and employment mechanism will lead to even more ferocious competition and the struggle for survival will intensify, with disequilibria and inequalities increasing, bringing about even more unemployment: an additional vicious circle will come into being (Bortis 2019, pp. 197–201). Given all this, consciousness about the absolute necessity of a sustainable economy might gradually fade away to a large extent as is indeed the case with leading politicians of certain countries and the already considerable number of climate sceptics might increase.

However, applying computer technology in a socially appropriate way will evidently increase the Common Good; for example, boring work at the assembly line (pictured in Charlie Chaplin's *Modern Times*) would be carried out by robots. However, creative work should always be done by human beings because computer software of some kind will never be able to recognise what is of fundamental and what is of secondary importance to perceive beauty and goodness, to be aware of problems and to propose solutions adapted to specific circumstances. Moreover, for the artisan, for example, producing a good is not just an economic and technical matter.

Producing a good as perfectly as possible provides an immense satisfaction and greatly contributes to the prospering of producers, through developing their dispositions and applying enhanced capacities acquired through professional practice.

However, producing goods predominantly through software applications leads on straightaway to a more unequal society and worsens the social situation as well as the situation of the social individuals in that boredom and frustration sets in. Economically, the few who develop the software will be well paid, those who apply it will, as a rule, be badly paid and those who own the software will, as a rule, become very rich; in fact, under this perspective, the basic aim of introducing computer software in all domains is not to promote the Common Good, but simply to *make money*. Moreover, as is very likely, involuntary unemployment will increase dramatically, and, as a consequence, the pressure on wages will grow, reducing effective demand which, in turn, will cause involuntary unemployment to increase further: a vicious circle comes into existence. As a consequence, chaotic situations may come into being in which the very rich, including the owners of software, will exercise not only economic and financial power but also political power. These socially destructive effects are bound to increase alienation, individual and social. Given this, *power* will ever more replace *Common Good oriented ethics* as the dominating force in "modern" societies. Here, we should remember that, in a Common Good perspective, the economy and technology should not dominate man (and society) but should stand in the service of man and of society, implying that not everything that is technologically feasible should be applied. The only sphere where all the great nations must be technologically at a top level is defence (military) technology; this is unfortunate, but nothing else than *Bismarckian Realpolitik.*

Finally, in a sustainable economy, small and medium-sized enterprises should dominate. For large enterprises whether privately or publicly owned, environmental regulations may be required. In short, a mixed economy will have to be built up. Materialism-cum-consumerism and profit and moneymaking have to be replaced by a strong impetus to aim at realising the Common Good which has multiple dimensions: social, political, cultural and ethical; however, the Common Good also implies organising production in industry and agriculture such that a harmonious relationship with nature comes into being.

8 Environmental Policies: Employment Projects and the Natural Environment

The second Great Transformation also implies realising Naomi Klein's Green New Deal (Klein 2019), which is also associated to Timothy Wise's call to promote family agriculture (Wise 2019). This transformation will be associated with important structural changes, implying that structural unemployment will be important during the transition period. Given this, permanent full employment policies will be of the utmost importance.

In a small article published in the middle of the Second World War (Keynes 1980/1942), Keynes proposes to increase employment through great projects, for example, "to rebuild London" (p. 264). An eminent architect objected: "Where is the money to come from?" (p. 264). Keynes replied that "we build houses with bricks and mortar, not with money" (p. 265). Indeed, money is only a device to mobilise real resources. And Keynes adds that the architect "was making the very usual confusion between the problem of finance for an individual and the problem for the community as a whole" (p. 266). In fact, "if we keep good employment when peace comes ..., even the post-war Budget problem will not be too difficult [also because of low interests in the future]" (p. 266). However, we must be allowed "to devote a large body of labour to capital works which would bring in no immediate return [this means increasing autonomous state or social expenditures (G) in relation (1) above]. Here is a real problem, fundamental but essentially simple, which is important for all of us to try to understand. *The [primary] task is to make sure that there is enough demand to provide employment for everyone*" (Keynes 1980/1942, p. 267, our emphasis). "To make sure of good employment we must have ready an ample programme of re-stocking and of development over a wide field, industrial, engineering, transport and agricultural—not merely building [and we may add, projects aiming at reducing environmental alienation through Naomi Klein's Green New Deal to reduce global warming or to change agricultural methods of production through promoting family farming as conceived by Timothy Wise, to decrease the use of pesticides for instance, are of crucial importance at present]" (p. 267). "Having prepared our blueprints covering the whole field of our requirements ... and these can be as ambitious and glorious as the minds of our engineers [in agriculture and industry] and social planners can conceive—those in charge must then concentrate on the vital task of central management, the *pace* at which the programme is put into operation, neither so slow as to cause unemployment nor too rapid as to cause inflation [or a deficit on current account]. The proportion of the [social] surplus which can be allocated to [reducing environmental alienation] must depend on the order of our preference between different types of project" (Keynes 1980/1942, pp. 267–68). It should be evident that in the present situation projects aimed at reducing environmental alienation, that is, projects aiming at realising a sustainable economy as they are implied, for example, by Klein (2019) and Wise (2019), are of primary importance. Making rivers, lakes and oceans clean again and reforestation are examples.

Keynes' suggestion that finance does not fundamentally matter is confirmed by the internal employment mechanism contained in the supermultiplier equation, that is, relation (1) above. Indeed, in macroeconomic equilibrium ($S = I$), the equation representing the internal output and employment mechanism becomes:

$$Q_i = (1/t)G, \tag{3}$$

where (t) is the tax–income ratio (see also Bortis 2019, p. 197). Government expenditures may increase through realising Keynes' projects just mentioned until full employment is reached. Of course, in the present situation, projects aiming at reducing environmental alienation ought to be given priority. This means moving towards a sustainable economy.

This economic meaning of relation (3) emerges more clearly if written as

$$G = tQ. \tag{3a}$$

This relation means that, in a monetary production economy, government expenditures (G) set the economy into motion if the internal mechanism prevails. A macroeconomic equilibrium is reached once a budget equilibrium is realised. In a way, if there is involuntary unemployment, increasing government expenditures bring about the tax revenues required to keep the state budget in equilibrium; this is analogous to Keynes' proposition that investment will generate the savings required to bring about a macroeconomic equilibrium. And the budget equilibrium will persist if (G) is raised through implementing environmental projects (relation 3a). Given this, finance emerges to be a disequilibrium phenomenon, required to finance the setting into motion of new projects which would lead to a temporary budget deficit. But this task could be taken on by Central Banks without any problem.

9 Concluding Remarks: Unleashing the "Social" Through Striving After the Common Good

The theoretical and technical aspects of setting up projects in the environmental domain in order to move in the direction of full employment and a sustainable economy are certainly of the greatest importance. Even more important, however, is to create a spirit of enthusiasm which would constitute a powerful engine to move towards a sustainable economy. To create this spirit an unleashing of the "social" through teamwork is required. We can see the "social" at work with football (soccer) teams and appreciate the immense strength the "social" unfolds in order to reach a social aim through participating in a team; every player gives everything for the team, restrained only by the rules of fair-play. As members of a team, the social individuals are complementary; co-operation and co-ordination are required to operate successfully.

Now one may conceive of society as a huge team in which the various parts of society—production of goods, cultural and social activities, the political and legal domain, for instance—are all complementary in view of reaching *one* common aim, that is, striving for the Common Good, which includes a great variety of components, for example, full employment and a sustainable economy, that is, harmony between man (society) and nature, with environmental alienation eliminated. As Keynes has perceived with greatest clarity, full employment is the most important component of the Common Good. Indeed, if politicians and social leaders tell (and show) to all those who feel excluded from society because of prolonged involuntary unemployment, to those living in precarious conditions, to those who live in poverty and even misery, that to reach the Common Good everybody is needed, nobody is to be excluded, then an immense hope and an equally immense enthusiasm will be created. *The entire social climate will change fundamentally.* Certainly, Keynes had this in mind when, in the middle of a terrifying World War, he wrote his short notes on creating full employment through realising big projects for the after-war peace period (Keynes 1980/1942, pp. 264–70). And Keynes also held that full employment could be reached in a socially appropriate way through the internal employment mechanism hinted at above and in

Bortis (1997 and 2019). Finally, Keynes was convinced that the internal employment mechanism could only be applied properly if the balance on current account could be kept in broad equilibrium in the long run and if there was some control of international capital movements (for a very short presentation of this issue, see Bortis 2019); this would require the setting up of a supranational money, the *Bancor*, and of a *Clearing Union* (Keynes 1980, *Activities 1940–1946*). In a way, this volume and the *General Theory* are complementary: In fact, the *General Theory* (Keynes 1973/1936) deals with the internal employment mechanism which operates within an economy and Keynes (1980, *Activities 1940–1946*) states that the economy considered in the *General Theory* need not be a closed and self-sufficient, but that mutually beneficial trade relations are possible to any extent, the only condition being that the balance on current account is in broad equilibrium in the long run (see also Bortis 2019, pp. 198, 205–6). Hence Keynes' overall message remains fully valid at the outset of the twenty-first century, also in relation with the issue of a sustainable economy.

References

Bortis, H. (1997), *Institutions, Behaviour and Economic Theory—A Contribution to Classical—Keynesian Political Economy*, Cambridge University Press, Cambridge, UK.

Bortis, H. (2019), "Principles underlying classical-Keynesian employment and distribution policies", in *A Modern Guide to State Intervention*, edited by Nikolaos Karagiannis and John E. King, Edward Elgar, Cheltenham, UK, and Northampton, MA, USA, pp. 191–209.

Daneke, G.A. (2020), "Machina-economicus or homo-complexicus: Artificial intelligence and the future of economics?", *Real World Economics Review*, issue no. 93, pp. 18–39.

Keynes, J.M. (1973/1936), *The General Theory of Employment, Interest and Money*. CW, vol. VII, Macmillan, London.

Keynes, J.M. (1980/1942), *Employment Policy—How Much Does Finance Matter?* CW, vol. XXVII, *Activities 1940–1946—Shaping the Post-War World—Employment and Commodities*, The Royal Economic Society (Macmillan and Cambridge University Press), London et al., pp. 264–70; orig. *The Listener*, 2 April 1942.

Keynes, J.M. (1980), *Activities 1940-1944, Shaping the Post-War World: The Clearing Union*, CW, vol. XXV, Macmillan and Cambridge University Press, London and Basingstoke.

Klein, N. (2019), *Warum nur ein Green New Deal unseren Planeten retten kann*, Hoffmann und Campe Verlag, Hamburg; American original 2019: *On Fire. The (Burning) Case for a Green New Deal.*

Mini, P. (1991), *Keynes, Bloomsbury and The General Theory*, Macmillan, London.

O'Donnell, R.M. (1989), *Keynes: Philosophy, Economics and Politics—The Philosophical Foundations of Keynes's Thought and their Influence on his Economics and Politics*, Macmillan , London.

Sennett, R. (2008), *Handwerk*, Berlin Verlag GmbH, Berlin; American original 2008: *The Craftsman*, Yale University Press, New Haven and London.

Wise, T. (2019), *Eating Tomorrow: Agribusiness, Family Farmers, and the Battle for the Future of Food*, The New Press, New York City.

Green New Deal Policies and the Decarbonisation Challenge

Richard Lewney

1 Introduction

Just as the crisis of the Great Depression motivated the major engagement of government in the Roosevelt New Deal, the climate change crisis has motivated calls for a bold mobilisation of the resources of government in a Green New Deal. At the same time, growing concern over the injustice of sustained inequality and the huge economic impact of the Covid-19 pandemic have prompted the inclusion of these additional objectives in Green New Deal or Green Recovery policy programmes.

This chapter begins by setting the context for these kinds of programmes and notes how their focus has developed in response to events over the past fifteen years. It then sets out the case for a post-Keynesian approach to modelling the impacts of the policies and illustrates this with two case study applications using the E3ME[1] global macro-sectoral model.

[1] www.e3me.com.

R. Lewney (✉)
Cambridge Econometrics, Cambridge, UK
e-mail: rl@camecon.com

N. Karagiannis and J. E. King (eds.), *Visions and Strategies for a Sustainable Economy*, Global Institute for Sustainable Prosperity,
https://doi.org/10.1007/978-3-031-06493-7_10

2 Responding to the Challenge of Global Warming

The Paris Agreement sets the objective of

> holding the increase in the global average temperature to well below 2 °C above pre-industrial levels and pursuing efforts to limit the temperature increase to 1.5 °C above pre-industrial levels. (Article 2 of UNFCCC 2015)

What is striking about this objective is how much more ambitious it is than the recommendation coming from the most influential Integrated Assessment Model,[2] DICE, the latest version of which is documented in Nordhaus (2018). Nordhaus (2018) reports[3] a projected global temperature rise by 2100 of 4.1 °C under a baseline 'no policy' case and an increase of 3.5 °C under a case in which the 'optimal' amount of abatement of emissions is undertaken. The concept of 'optimality' trades off the GDP deemed to be lost when investment is diverted from other kinds of productive use to carbon emissions abatement with the GDP that will be lost due to the damages associated with unabated temperature increase, discounting future costs and benefits at a rate intended to reflect society's preference for consumption today over consumption tomorrow. The conclusion that an increase of 3.5 °C would be optimal stands in marked contrast to the language of the Intergovernmental Panel on Climate Change:

> Without additional mitigation efforts beyond those in place today, and even with adaptation, warming by the end of the 21st century will lead to high to very high risk of severe, widespread and irreversible impacts globally (high confidence). Mitigation involves some level of co-benefits and of risks due to adverse side effects, but these risks do not involve the same possibility of severe, widespread and irreversible impacts as risks from climate change, increasing the benefits from near-term mitigation efforts. (IPCC 2014, p. 17)

[2] An 'Integrated Assessment Model' combines equations intended to represent the processes of the economy, greenhouse gas emissions and the cost of abatement measures, global warming and the damage to the economy that would arise under a future with higher global temperatures.

[3] Nordhaus (2018), Table 4, p. 353.

Lewney (2021) reviews the reasons why the conclusions from DICE may be biased against strong mitigation action; why, in effect, the perspective of an economist coming from the neoclassical tradition should differ so strongly from the scientific consensus represented by the IPCC. The scientists may be operating in a different ethical framework, placing a value on the earth's ecosystem that is not captured merely in its contribution to GDP. They may not accept the neoclassical assumption that the 'baseline' economy operates at an optimal production frontier so that more abatement necessarily entails a sacrifice of other investment or consumption. They may not accept that the welfare of future generations should be given less weight than that of those living now. But the main reason is likely to be a different assessment of the scale of damages that could arise from a temperature increase of more than 2 °C and, given the large uncertainties involved, the importance of precautionary action as insurance to mitigate potentially catastrophic risks (Keen 2020).

In any event, the outcome has been that mainstream economics has been of limited relevance as the focus of the policy debate has been on securing commitments in line with the Paris Agreement. The return of the US to the Paris Agreement in January 2021 and the announcement by President Xi Jinping in September 2020 that China will seek to achieve carbon neutrality by 2060 have given renewed impetus to the effort to strengthen global action. At the same time, growing confidence that these commitments will finally be translated into action has been reflected in the financial community. On the one hand, there has been a growing appetite among institutional investors for 'green' products to support shifts towards 'Paris-aligned' portfolios. On the other, there has been growing concern about the potential for a disruptive repricing of risk for fossil fuel assets since it is clear that the world's carbon budget cannot accommodate full exploitation of what had previously been regarded as economically recoverable reserves of coal, oil and gas. The landmark speech of Carney (2015) has been followed by a concerted movement of regulators of financial institutions to require stress testing and risk reporting of climate-related risks, a programme taken forward by the Network for Greening the Financial System.

Despite these encouraging signs of movement, the scale of the challenge to move to net zero greenhouse gas emissions over the period of just a few decades remains enormous. IPCC (2018) includes a review of alternative mitigation 'pathways': possible combinations of population change, economic growth, technology take up and lifestyle change that together provide a scenario consistent with 1.5 °C warming. While there is considerable uncertainty about the pathways, some broad ingredients are clearly essential: electricity generation from renewable sources; electrification of energy demand in buildings, transport and industry; investment in energy efficiency; widespread use of energy stores, including batteries, to match electricity supply from non-dispatchable renewable sources to daily and seasonal demand peaks; some form of carbon dioxide removal such as increased afforestation or bioenergy with carbon capture and storage; large reductions in GHG emissions from agriculture, perhaps supported by lifestyle and dietary changes. The proposals included in Green New Deal programmes therefore need to incorporate policies to stimulate the investment, innovation, financing and behaviour change to bring about this kind of future and economic modelling of those proposals needs to be capable of representing it.

3 Green New Deal and Green Recovery Programmes

To this author's knowledge, the term 'Green New Deal' was first coined by Thomas Friedman in an op-ed for the *New York Times* in January 2007 (Friedman 2007). His call for action at that time was motivated by the scale of the challenge required to address climate change and recognition of the economic opportunities offered. Just as Roosevelt's New Deal involved a wide range of projects, programmes, regulations and reforms to lead the US economy out of the Great Depression, so a wide range of policy actions were needed to free the energy system from its reliance on combustion of fossil fuels and the consequent carbon emissions. Friedman argued for strong government intervention in the form of both tough regulatory standards and higher prices to drive and incentivise behaviour change.

Friedman's article was published in the early stages of the subprime lending crisis and well before the spread of bankruptcies and consequent bank lending freeze had triggered the Great Recession. The parallel with Roosevelt's New Deal did not, therefore, extend to the requirement for a bold and expansionary fiscal policy to combat recession. But by 2008 the global economic crisis was fully apparent and the 'Green New Deal Group' in the UK was calling for a programme of actions to address both the short-term economic stabilisation needs of the recession and the long-term challenge of keeping global warming below 2 °C (Sims et al. 2008).

In the UK, the Group called for substantial publicly funded investment to improve the energy efficiency of homes, a drive to stimulate innovation and low-carbon technological change and a shift from an economy reliant on a large financial sector to one with a large number of jobs engaged in environmental transformation. Funding would come in part from carbon taxes to raise fossil fuel prices to better reflect the environmental externalities of carbon and other air emissions, a crackdown on tax evasion and from financial innovation to mobilise private capital. The vulnerability of low-income households to higher fuel prices was recognised to the extent that the policy recommendations included support to reduce energy consumption by improving the energy efficiency of their homes. Given the context of the financial crisis and the subsequent socialisation of the debts of private banks, the report included proposals for a substantially reformed financial system.

In the event, neither version of the Green New Deal gained much traction with governments. The loss of tax revenue during the recession and the cost of bank bail-outs left governments much more heavily indebted than before the recession. Climate change slipped among policy priorities and economic orthodoxy demanded fiscal austerity to reduce government debt. Within the European Union, some progress was made in agreeing tighter targets for cutting emissions, including regulation of road vehicle emissions, and the cap and trade Emissions Trading System launched in 2005 continued to be developed, but an over-generous allocation of free allowances kept the carbon price low and effectively exempted the most energy-intensive industries for the first decade of its operation. Even so, there was sufficient expansion of the global market for solar photovoltaic panels and wind turbines to trigger innovation that brought about a substantial reduction in the unit cost of installations.

The decade from 2010 saw most OECD economies recover albeit to slower rates of GDP growth than had been typical before the crisis. Labour productivity growth was also slow, so that unemployment rates fell, first in the US and Japan and later in Europe. Attention shifted from the short-term economic crisis to the long-term problem of slow-growing or stagnant living standards for low-income groups and growing inequality. This motivated a new concept of the Green New Deal in the US, embodied in Recognizing the duty of the Federal Government to create a Green New Deal (2019), introduced to Congress by Representative Alexandria Ocasio-Cortez and Senator Edward Markey. In this conception, the aim was to tackle the twin problems of climate change and inequality. The breadth of these proposals, and the strong role envisaged for the Federal Government, came in marked contrast to a counter resolution, the Energy Innovation and Carbon Dividend Act of 2019 (2019), introduced into Congress by Representative Ted Deutch in January 2019. This proposed an escalating carbon 'dividend' (i.e. a tax) levied on fossil fuels to be recycled in equal shares to individuals, accompanied by a 'carbon border fee adjustment' import tariff.

In the UK, the Labour Party included Green New Deal proposals in its November 2019 election manifesto (Labour Party 2019). The concern about energy poverty noted in Sims et al. (2008) was also evident here, with a commitment to bring the bulk of the UK's 27 million homes up to the highest energy-efficiency standards. But Labour lost the election and so the manifesto never became government policy.

Whereas, so far, none of the initiatives described above have actually been implemented by a government, the European Union is proceeding with its European Green Deal (European Commission 2019). This comprises a number of measures, mostly focused on reducing greenhouse gas emissions but also including certain other environmental priorities (such as supporting biodiversity and promoting a circular economy). It envisages €1 trillion to be spent on sustainable investments over the coming decade, half of which would come from the EU budget (with the rest to come from national governments and the private sector). The European Green Deal also includes the Just Transition Mechanism, a package of €140bn over ten years to support workers and communities in regions most exposed to transition risk (notably the locations where fossil fuel is extracted and processed). But it does not include the ambition to tackle the underlying causes of the long-term trend towards greater inequality expressed in Recognizing the duty of the Federal Government to create a Green New Deal (2019).

The Covid-19 pandemic that took hold in 2020 shifted the policy focus again. The initial impact was dominated by the supply and demand shocks associated with the lockdown policies that largely prevented social consumption activities, extensive travel and face-to-face working. But the growing availability of vaccines in 2021 offered the prospect of an end to these restrictions and a 'return to normal'. The question then was: what kind of normal?

The slogan 'build back better', originally developed to promote a response to natural disasters that goes beyond mere replacement of the damaged physical assets (Clinton 2006), has been adopted by various groups advocating 'Green Recovery' measures of one kind or another, among them WBCSD (2020), Biden for President (2020), We Mean Business (2020) and the UK campaigning group Build Back Better.[4] These programmes recognise the short-term need for macroeconomic demand stimulus measures as spending by households and firms is constrained by the deterioration in balance sheets experienced during 2020. But they call for that stimulus to align with long-term structural goals: decarbonising the energy system, creating high-quality jobs and addressing other social objectives. This reflects a recognition that the pandemic's economic effects are felt very unevenly across households and firms, with the prospect of widely different outcomes in a 'K-shaped' recovery (Bheemaiah et al. 2020).

4 A Post-Keynesian Approach to Modelling the Economic and Environmental Impact of a Green Recovery Programme

4.1 Motivation

The challenge of designing policies to promote short-term recovery from the Covid-19 pandemic and long-term climate change mitigation has features that make post-Keynesian insights particularly relevant.

Firstly, of course, economic conditions currently show broadly similar features to those of the Great Depression when Keynes was arguing for fiscal stimulus. Private sector spending is weak, and there is little prospect that monetary policy can alter that: policy interest rates are already very low and the money injected by unconventional monetary policy is supporting asset prices rather than spending on goods and services. Because the crisis is a global one, no country can rely on demand for its exports to lead recovery. There is substantial excess capacity, inflation is low and the idea that a full-employment equilibrium could be restored by reductions in wage rates is not credible. The relaxation of social mixing restrictions allows some recovery in activity from the record falls seen in 2020 as some pent-up demand is released, but many firms and households have entered the recovery with less wealth and greater indebtedness than before the pandemic. Widespread predictions of a rapid recovery were confounded following the 2008 recession (Lewis and Pain

[4] See buildbackbetteruk.org.

2015), consistent with Keynesian scepticism about the efficacy of strong self-equilibrating processes to drive recovery following a deep recession.

Secondly, a post-Keynesian understanding of the endogenous process of money creation supports the view that there is little likelihood that a loosening of fiscal policy would crowd out private spending (or cause an inflationary spiral). Investment spending is not constrained by a shortage of saving and 'loanable funds', but by the willingness of firms to take on additional debt and expand capacity at a time when demand is weak and losses have weakened balance sheets.

Thirdly, present circumstances seem to be better described by the notion that households and firms adapt their behaviour to cope with conditions of radical uncertainty, in the post-Keynesian tradition, rather than operate on the basis of a well-defined probability distribution of possible future outcomes. This is particularly relevant when we consider whether and how to stimulate take-up of less familiar, low-carbon technologies on a pathway towards net zero emissions.

Two requirements for effective analysis of the impact of Green Recovery policies are less strongly represented in the post-Keynesian tradition. The first is disaggregation of macroeconomic aggregates to distinguish sectors and products with different characteristics. The second is an endogenous treatment of innovation.

With regard to disaggregation, post-Keynesians recognise the principle of heterogeneity of actors, and reject the 'representative agent' device often used to justify the application of microfoundations derived from neoclassical theory to model macroeconomic aggregates, but post-Keynesian models have not typically explicitly distinguished groups of firms or households. For the analysis of climate change mitigation pathways, however, important insights are lost if no distinction is made between energy-intensive and non-energy-intensive sectors, or between fossil fuel-based and renewable energy sources, or between the technologies required to support the low-carbon economy and those which extract, process and burn fossil fuels. In the applied modelling reported below, a macro-sectoral model that makes these distinctions is used. It may equally be said that to analyse policies to address economic inequality and promote a Just Transition, the model should distinguish different types of labour and households and not simply wage incomes and profits. The model used below distinguishes jobs in different sectors, but not different types of labour; however, it does split household incomes and consumption into quintiles so that some insight into impacts on inequality is possible.

With regard to endogenous innovation, the importance to climate change mitigation of the large reductions in the cost of key renewable electricity technologies seen over the past decade or more cannot be overstated (IRENA 2020). There is a large literature on the economics of environmental innovation and the potential role of environmental policy to stimulate innovation (Popp et al. 2010; Popp 2019) which in turn promotes cost reductions and greater take-up. Mercure et al. (2016) trace the way in which different schools in economics have sought to incorporate innovation, distinguishing the contributions of those coming from the neoclassical equilibrium school and those that follow a non-equilibrium approach. Mercure et al. group together as non-equilibrium approaches both Schumpeter and those following Keynes, notably Kaldor, arguing for complementarity of these two traditions. Both emphasised the role of profit-seeking entrepreneurs driving innovation, the key part played by credit creation and finance to support that innovation and the real incomes of entrepreneurs and, depending on the extent of competition, consumers as the benefits of innovation are passed on in lower prices. The model applied in the case studies in this chapter incorporates a treatment of diffusion of new technologies based on this Schumpeterian/Kaldorian view of innovation.

4.2 The E3ME Model

E3ME is a macro-sectoral econometric model of the global economy in the post-Keynesian tradition distinguishing some 60 countries and regions. It was first developed in the 1990s through the European Union's research framework programmes with the explicit aim of quantifying the economic impacts of climate change mitigation policies. It has a detailed treatment of the energy system, distinguishing different kinds of energy products and, for the most important sources of carbon emissions, different kinds of energy transformation technologies. It includes a range of policy instruments for promoting climate change mitigation including, but not limited to, a carbon price instrument. In line with post-Keynesian theory, E3ME does not assume that prices adjust to clear markets; rather, prices are determined as a mark-up on costs and supply and demand mismatches are mainly resolved through supply quantity (output) adjustments. Consequently, it is possible in E3ME for lower spending to drive lower output, incomes and investment and for output to be lost permanently in a recession.

Many of the key behavioural economic relationships are estimated as time-series econometric equations in the tradition of structural economic models (Wren-Lewis 2018). There are over 30 sets of econometrically estimated equations, including the components of GDP (consumption, investment, international trade), prices, energy demand and materials demand. Each equation set is disaggregated by country and by sector. Responses to policy changes are therefore based on historically observed relationships between variables. Like any such econometric relationships, the results depend on the validity of the estimated parameters in the simulation or projection being undertaken and hence are potentially vulnerable to the 'Lucas critique'. But, of course, the approach in DSGE modelling of imposing strong neoclassical assumptions about household and firm behaviour only succeeds if those untested assumptions hold; the fact that such models typically fit the data very poorly (Wren-Lewis 2018) is not encouraging in this respect.

For key energy transformation technologies (e.g. power generation, road transport), the choice of technology for new investment is modelled using a method of cross-sectional discrete choice modelling known as Future Technology Transformations, or FTT (Mercure, Pollitt, Edwards, et al. 2018; Mercure, Pollitt, Viñuales, et al. 2018; Mercure et al. 2019), which is intended to represent the Schumpeterian/Kaldorian view of innovation and growth noted earlier. Among its physical indicators relevant to environmental pressures, E3ME includes carbon emissions, emissions of other (local) air pollutants and the use of certain raw material resources. E3ME does not include a module to estimate the global warming implications of alternative pathways for greenhouse gas emissions, but it can be linked to a separate carbon cycle and atmosphere circulation model of intermediate complexity (GENIE), a brief description of which is included in Mercure, Pollitt, Edwards, et al. (2018).

E3ME is a simulation model rather than an optimisation model. That is, it is intended to simulate how firms and households actually behave rather than to find an optimal pathway to achieve any given set of objectives. Given that optimising behaviour is not assumed or imposed, the model outcomes are not interpreted as representing outcomes that are optimal either for individual agents or for society as a whole. Instead, a range of indicators (e.g. GDP, employment, unemployment, carbon emissions) are produced to inform an assessment of whether one scenario is preferable to another, in much the same way that these indicators are available historically for an assessment of past performance on economic, social and environmental criteria. Precisely because optimising behaviour is not assumed in the baseline, it is possible for alternative policies to produce scenarios in which not only is environmental degradation less (that is, damaging externalities are reduced) but incomes are higher (private outcomes are better).

E3ME has been used widely to simulate the economic impacts of climate change mitigation strategies. In addition to the case studies described later in this chapter, examples include Barker et al. (2016), European Commission (2018), IRENA (2018), Lee et al. (2018), Mercure, Pollitt, Viñuales, et al. (2018), The Global Commission on the Economy and Climate (2018) and Pollitt et al. (2014).

4.3 Using E3ME to Model a Green Recovery

We summarise here the approach taken and results found for two recent case studies that apply E3ME to model a Green Recovery programme.

In the first case study, Cambridge Econometrics (2020) constructed scenarios with annual results to 2030 for a 'baseline' projection, incorporating the impacts of the Covid-19 crisis but with no recovery programme and for two alternative recovery programmes. The two programmes were designed to require, ex ante, the same size of increase in government spending and assumed that in both cases this would be accommodated by an increase in government debt.[5] One 'conventional' programme provided a generalised boost to household spending by introducing a temporary reduction in consumption taxes (VAT or sales tax, depending on the tax system operating in the country), a policy that was followed in some countries during the 2008 financial crisis. In the modelling a cut of 5 percentage points from baseline levels over 2021–2023 was assumed, followed by gradual increases in the tax rate back to baseline levels

[5] Note that the *difference in outcome between the two recovery scenarios* would be broadly similar whatever mechanism was used to finance it, but the extent of recovery, represented by the *difference from the Covid-19 baseline*, would be smaller for both scenarios in E3ME if they were funded by tax increases or spending cuts designed to prevent an increase in budget deficits.

by 2028. The alternative programme, for a Green Recovery, included measures targeting fossil fuel use so as to further reduce carbon emissions: public investment to improve the energy efficiency of buildings, subsidies for wind and solar power, public investment to upgrade electricity grids to support more extensive use of intermittent renewables, a car scrappage scheme for purchasers of electric vehicles and a tree-planting programme. Assumptions for the scale of these measures were based on an assessment of what might be plausible, and then cuts in consumption taxes were sized and added to the package to bring the total ex ante cost to the same value as the conventional programme.

The impacts of the scenarios at global level on GDP and jobs are shown in Figs. 1 and 2.

The figures show differences from a projection of the outcomes that would have obtained if the pandemic had not occurred. Figure 2 shows the 8% estimate of the fall in GDP in 2020. Thereafter, some recovery is projected and the Covid-19 baseline has faster growth than was expected in the no-pandemic case (the negative impact compared with no-Covid-19 baseline is reduced over time), but some output is lost permanently, reflecting in particular the impact of the crisis on investment. The boost to spending in the VAT Recovery scenario leads to a stronger outcome in 2021 but the phasing out of the tax cut stimulus over time leaves GDP at about the same level as in the Covid-19 baseline by the end of the decade.

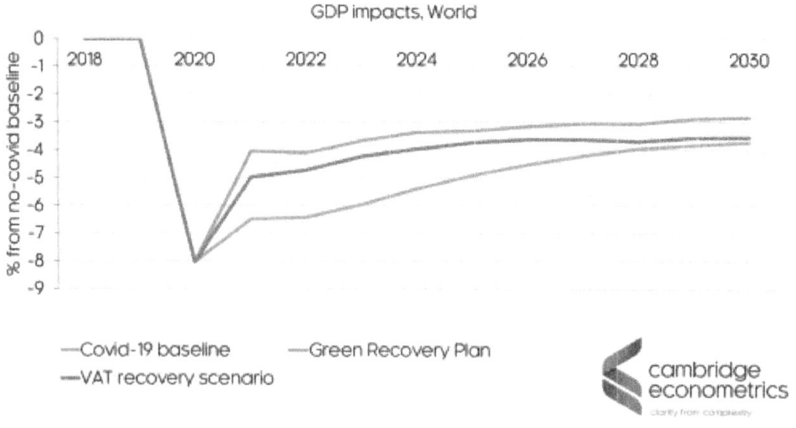

Fig. 1 Global employment impacts in non-green and green recovery scenarios

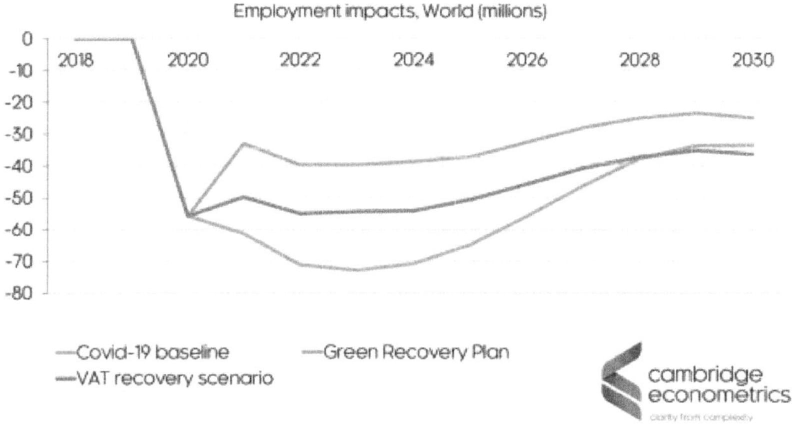

Fig. 2 Global GDP impacts in non-green and green recovery scenarios

The Green Recovery scenario produces a stronger outcome for GDP than the VAT Recovery, by about 0.5–1% of no-Covid GDP in any given year. This comes about because decarbonisation technologies substitute capital for ongoing consumption of fossil fuels and so the push towards greater decarbonisation levers in more private investment. In addition, the green economy activities that expand in the Green Recovery scenario tend to be somewhat more labour-intensive than the fossil fuel extraction and processing that contract, so that there is a net gain in jobs as shown in Fig. 1. E3ME assumes that the propensity to consume is higher for wage incomes than for profits, and so household consumption is also higher. There is spare capacity in the economy at the beginning of the period so that higher spending leads to higher output rather than higher inflation, and E3ME has the property that investment responds to a reduction in spare capacity: hence, provided that the acceleration is not so fast as to outpace the expansion of capacity, higher spending can be met by higher production in the long term also. Results for any given country depend on the extent to which its economy specialises in fossil fuel extraction and processing and the extent to which it is likely to be an importer or exporter of green technologies.

The results for GDP and jobs show that, for the same ex ante fiscal injection, better outcomes can be achieved with a package of green recovery measures than with one that focuses on consumption tax cuts.

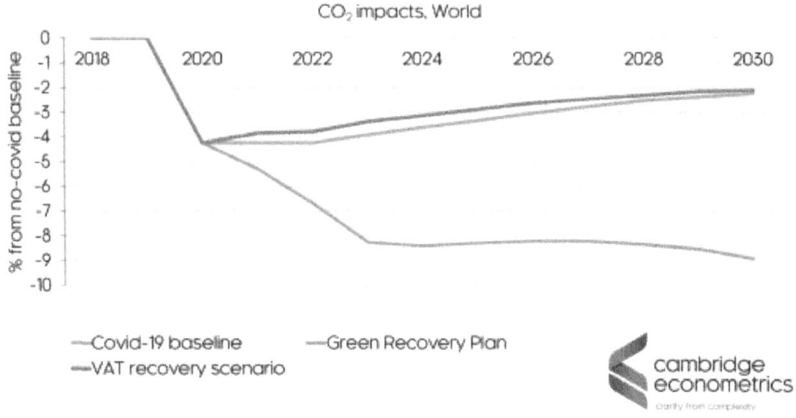

Fig. 3 Global CO_2 impacts in non-green and green recovery scenarios

However, the difference is not large for these economic indicators. What is very different, of course, is the outcome for carbon emissions shown in Fig. 3.

The Covid-19 baseline sees a sharp fall in carbon emissions in 2020 but then a gradual return towards the no-Covid level of emissions by 2030. The VAT Recovery sees the same profile for emissions but slightly higher, reflecting the higher level of economic activity and no measures to offset the consequent boost to fossil fuel energy consumption and emissions. In contrast, the Green Recovery sees a continued fall in emissions over 2021–2023 to a level more than 8% below the no-Covid level which is then sustained over the rest of the period despite the closure of the gap for GDP. While the scale of these measures is by no means sufficient to assure a pathway consistent with the goals of the Paris Agreement, they do make a significant contribution to reducing emissions.

In the second case study, ECLAC (2020) used E3ME[6] to simulate two levels of ambition of green recovery. In the study's 'Unconditional Big Push' scenario, the countries of Latin America and the Caribbean introduce a recovery package with policies that target reductions in greenhouse gas emissions in line with their unconditional commitments under their

[6] ECLAC staff developed the scenarios using E3ME under licence with technical support from Cambridge Econometrics.

Nationally Determined Contributions (implying a 13% cut in emissions relative to a Covid-19 baseline by 2030); only the unconditional targets are met because of the assumption that the rest of the world takes no accompanying action. In the 'Conditional Big Push' scenario, the rest of the world introduces green fiscal reforms and provides support to Latin America and the Caribbean to promote reforestation, allowing them to meet the more ambitions targets of their conditional NDCs (a 23% cut by 2030).

To achieve the unconditional target reduction in GHG emissions in the Unconditional Big Push scenario, ECLAC (2020) assumed that governments introduce subsidies to kick-start take-up of non-conventional renewable energy technologies in power generation and extend regulations to promote a higher biofuel blend and adoption of low-emission road transport vehicles. ECLAC (2020) went further than Cambridge Econometrics (2020) in also including policies to address social inequality: an increase in public spending on health to support better access to medical care among the poor. ECLAC (2020) also differed from Cambridge Econometrics (2020) in assuming budget neutrality compared with the Covid-19 baseline: the study assumed that fossil fuel subsidies would be phased out to finance the other measures. In the Conditional Big Push scenario, the study assumed an across-the-board green fiscal reform in the rest of the world, raising a tax of €27 per tonne of CO_2 and using the receipts to lower VAT and payroll taxes, together with support reforestation programmes in Latin America and the Caribbean extending to 3.8 m ha over 2020–2030 and hence cutting the region's net GHG emissions.

By design, the two scenarios achieve substantial reductions in GHG emissions. The economic impacts in the two scenarios are similar: an increase of about 2.3% in GDP and 0.7% in employment over the Covid-19 baseline by 2030. Energy costs are lower in the Unconditional Big Push scenario (thanks to lower renewable technology costs in a larger global market and lower firewood costs due to expanded forests) and this improves the real incomes particularly of poorer households for whom energy costs form a larger share of consumption spending.

5 Conclusions

Mainstream neoclassical economics underestimates the risks to GDP and wellbeing from global warming and overestimates the cost of mitigation action. Its recommendations for action are too modest and have effectively been ignored by the policy consensus represented by the Paris Agreement. Similarly, macroeconomic models based on neoclassical assumptions are not well suited to the analysis of the twin problems of entrenched inequality and the impact of the Covid-19 pandemic. Climate mitigation policy design is better served by a fusion of post-Keynesian macroeconomics with Schumpeterian innovation theory, combined with disaggregation of sectors, products and technologies to distinguish carbon-intensive and low-carbon activities. Post-Keynesian scepticism about strong self-equilibrating macroeconomic processes makes that approach relevant also to an assessment of the impact of macroeconomic recovery packages that also target green objectives.

This approach has been implemented in a well-established, applied macro-sectoral model of the global economy. Economic modelling of Green New Deal and Green Recovery policies within that model generally shows that those policies are effective in stimulating GDP and jobs. The model's macro-sectoral features highlight the radical shift in economic structure, with the closure of fossil fuel extraction and processing and expansion of sectors producing the technologies to support electricity generation from renewable sources, energy storage and energy efficiency. The results highlight the importance of Just Transition policies to anticipate these changes and support workers and communities that are dependent on the production of fossil fuels, and to mitigate the impact of the higher prices of fossil fuels that will be necessary as one of the policies to incentivise the transition.

References

Barker, T. S., Alexandri, E., Mercure, J.-F., Ogawa, Y., and Pollitt, H. 2016. "GDP and employment effects of policies to close the 2020 emissions gap." *Climate Policy*, Volume 16, Issue 4, pp. 393–414.

Bheemaiah, K., Esposito, M., and Tse, T. 2020. "Are we experiencing a K shaped recovery from COVID-19?" World Economic Forum. Available at https://www.weforum.org/agenda/2020/12/k-shaped-covid19-coronavirus-recovery/.

Biden for President. 2020. "Build back better: Joe Biden's jobs and economic recovery plan for working families." Policy statement by the Biden/Harris Election Campaign. Available at https://joebiden.com/build-back-better/.

Cambridge Econometrics. 2020. "Assessment of green recovery plans after COVID-19, we mean business coalition." Available at https://www.wemeanbusinesscoalition.org/wp-content/uploads/2020/10/Green-Recovery-Assessment-v2.pdf.

Carney, M. 2015. "Breaking the tragedy of the horizon—Climate change and financial stability." Speech given to Lloyd's of London, Bank of England, London. Available at https://www.bankofengland.co.uk/-/media/boe/files/speech/2015/breaking-the-tragedy-of-the-horizon-climate-change-and-financial-stability.pdf?la=en&hash=7C67E785651862457D99511147C7424FF5EA0C1A, 29 9.

Clinton, W. J. 2006. "Key propositions for building back better: Lessons learned from tsunami recovery." Office of the Secretary-General's Special Envoy for Tsunami Recovery, United Nations, New York. Available at https://www.recoveryplatform.org/assets/publication/Tsunami%20Reocvery/key%20propositions%20for%20building%20back%20better%20-%20tsunami%20-%20UN.pdf.

Economic Commission for Latin America and the Caribbean (ECLAC). 2020. "Building a new future: Transformative recovery with equality and sustainability (LC/SES.38/3-P/Rev.1)." Santiago. Available at https://repositorio.cepal.org/bitstream/handle/11362/46228/4/S2000698_en.pf.

Energy Innovation and Carbon Dividend Act of 2019. 2019. H.763, 116th Cong. Available at https://www.congress.gov/bill/116th-congress/house-bill/763.

European Commission. 2018. "In-depth analysis in support of the commission communication COM(2018) 773." *A clean planet for all a European long-term strategic vision for a prosperous, modern, competitive and climate neutral economy*. Available at https://ec.europa.eu/clima/sites/clima/files/docs/pages/com_2018_733_analysis_in_support_en_0.pdf.

European Commission. 2019. "The European Green Deal", Communication from the Commission to the European Parliament, the European Council, the Council, the European Economic and Social Committee and the Committee of the Regions, COM(2019) 640 final, 11.12.2019, Brussels. Available at https://eur-lex.europa.eu/resource.html?uri=cellar:b828d165-1c22-11ea-8c1f-01aa75ed71a1.0002.02/DOC_1&format=PDF.

Friedman, T. L. 2007. "A warning from the garden." *New York Times*, 19 January 2007. Available at https://www.nytimes.com/2007/01/19/opinion/19friedman.html.

IPCC. 2014. "Summary for policy-makers." *AR5 synthesis report: Climate change 2014*. IPCC. Available at https://www.ipcc.ch/site/assets/uploads/2018/02/SYR_AR5_FINAL_full.pdf.

IPCC. 2018. *Global Warming of 1.5 °C. An IPCC Special Report on the impacts of global warming of 1.5 °C above pre-industrial levels and related global greenhouse gas emission pathways, in the context of strengthening the global response to the threat of climate change*. Intergovernmental Panel on Climate Change.

IRENA. 2018. "Global energy transformation: A roadmap to 2050." International Renewable Energy Agency, Abu Dhabi. Available at https://www.irena.org/-/media/Files/IRENA/Agency/Publication/2018/Apr/IRENA_Report_GET_2018.pdf.

IRENA. 2020. "Renewable power generation costs in 2019." International Renewable Energy Agency, Abu Dhabi. Available at https://www.irena.org/-/media/Files/IRENA/Agency/Publication/2020/Jun/IRENA_Power_Generation_Costs_2019.pdf.

Keen, S. 2020. "The appallingly bad neoclassical economics of climate change." *Globalizations*. https://doi.org/10.1080/14747731.2020.1807856.

Labour Party. 2019. "A Green industrial revolution." *The Labour Party 2019 Manifesto*. Labour Party, UK. Available at https://labour.org.uk/manifesto/a-green-industrial-revolution/.

Lee, S., Chewpreecha, U., Pollitt, H., and Kojima, S. 2018. "An economic assessment of carbon tax reform to meet Japan's NDC target under different nuclear assumptions using the E3ME model." *Environmental Economics and Policy Studies*, Volume 20, Issue 2, pp. 411–429.

Lewis, C., and Pain, N. 2015. "Lessons from OECD forecasts during and after the financial crisis." *OECD Journal: Economic Studies*. OECD. Available at https://www.oecd.org/economy/growth/Lessons-from-OECD-forecasts-during-and-after-the-financial-crisis-OECD-Journal-Economic-Studies-2014.pdf.

Lewney, R. 2021. "Environmental policies to save the planet." In *Economic policies for a post-neoliberal world*, Philip Arestis and Malcolm Sawyer, eds. Palgrave Macmillan, pp. 179–223.

Mercure, J.-F., Knobloch, F., Pollitt, H., Lewney, R., Rademakers, K., Eichler, L., van der Laan, J., and Paroussos, L. 2016. *Policy-induced energy technological innovation and finance for low-carbon economic growth*. Study on the Macroeconomics of Energy and Climate Policies, European Commission, Directorate-General for Energy. Available at https://ec.europa.eu/energy/sites/default/files/documents/ENER%20Macro-Energy_Innovation_D2%20Final%20%28Ares%20registered%29.pdf.

Mercure, J.-F., Knobloch, F., Pollitt, H., Paroussos, L., Scriecu, S., and Lewney, R. 2019. "Modelling innovation and the macroeconomics of low-carbon transitions: Theory, perspectives and practical use." *Climate Policy*, Volume 19, Issue 8, pp. 1019–1037. Available at https://www.tandfonline.com/doi/full/10.1080/14693062.2019.1617665?src=recsys.

Mercure, J.-F., Pollitt, H., Edwards, N. R., Holden, P. B., Chewpreecha, U., Salas, P., Lam, A., Knobloch, F., and Viñuales, J. E. 2018. "Environmental impact assessment for climate change policy with the simulation-based integrated assessment model E3ME-FTT-GENIE." *Energy Strategy Reviews*, Volume 20, April, pp. 195–208.

Mercure, J.-F., Pollitt, H., Viñuales, J. E., Edwards, N. R., Holden, P. B., Chewpreecha, U., Salas, P., Sognnaes, I., Lam, A., and Knobloch, F. 2018. "Macroeconomic impact of stranded fossil fuel assets." *Nature Climate Change*, Volume 8, pp. 588–593.

Nordhaus, W. D. 2018. "Projections and uncertainties about climate change in an era of minimal climate policies." *American Economic Journal: Economic Policy*, pp. 333–360. Available at https://pubs.aeaweb.org/doi/pdfplus/10.1257/pol.20170046.

Pollitt, H., Alexandri, E., Chewpreecha, U., and Klaassen, G. 2014. "Macroeconomic analysis of the employment impacts of future EU climate policies." *Climate Policy*, Volume 15, Issue 5, pp. 604–625.

Popp, D. 2019. "Environmental policy and innovation: A decade of research." National Bureau of Economic Research Working Paper 25631, NBER. Available at https://www.nber.org/system/files/working_papers/w25631/w25631.pdf.

Popp, D., Newell, R., and Jaffe, A. B. 2010. "Energy, the environment, and technological change." In *Handbook of the economics of innovation*, vol. 2, Bronwyn Hall and Nathan Rosenberg, eds., Academic Press/Elsevier, pp. 873–937.

Recognizing the duty of the Federal Government to create a Green New Deal. 2019. H. 109, 116th Cong. Available at https://www.congress.gov/bill/116th-congress/house-resolution/109/text.

Sims, A., Pettifor, A., Lucas, C., Secrett, C., Hines, C., Legget, J., Elliot, L., Murphy, R., and Juniper, T. 2008. "A Green New Deal, first report of the Green New Deal Group, New Economics Foundation, London." Available at https://neweconomics.org/uploads/files/8f737ea195fe56db2f_xbm6ihwb1.pdf.

The Global Commission on the Economy and Climate. 2018. "Unlocking the Inclusive Growth Story of the 21st Century: Accelerating Climate Action in Urgent Times." New Climate Economy project. Available at https://newclimateeconomy.report/2018/.

UNFCCC. 2015. "Paris Agreement, UNFCCC, Bonn." Available at https://unfccc.int/process/conferences/pastconferences/paris-climate-change-conference-november-2015/paris-agreement.

WBCSD. 2020. "To build back better, we must reinvent capitalism. Here's how." World Economic Forum. Available at https://www.weforum.org/agenda/2020/07/to-build-back-better-we-must-reinvent-capitalism-heres-how/.

We Mean Business. 2020. "How to build back better after COVID-19." World Economic Forum. Available at https://www.weforum.org/agenda/2020/04/how-to-build-back-better-after-covid-19/.

Wren-Lewis, S. 2018. "Ending the microfoundations hegemony." *Oxford Review of Economic Policy*, Volume 34, Issue 1–2, pp. 55–69.

Sustainability and Area Studies

Building Sustainable Communities in the Global South: The Communitarian Revolutionary Subject

David Barkin

1 Introduction

In the context of the prevailing abundance of diversity (biological, ethnic), the profound social inequalities, and the trends and attitudes of hegemonic forces in Latin America, a coherent process of environmental governance is proving difficult and environmental injustice is aggravated. In virtually every country in the region, increasing subordination to the global market has led to dramatic transformations in productive structures and processes along with the oft-times violent opening of new territories to domestic and foreign investment in renewable energy projects, primary production for international markets, and natural resources exploitation. These changes are provoking direct confrontations between, on the one hand, domestic policy makers, well-financed investors positioned to operate in international markets, purveyors of technologies, investors with concessions in regions and sectors recently opened to foreign investment;

D. Barkin (✉)
Universidad Autónoma Metropolitana, Mexico City, Mexico
e-mail: barkin@correo.xoc.uam.mx

N. Karagiannis and J. E. King (eds.), *Visions and Strategies for a Sustainable Economy*, Global Institute for Sustainable Prosperity,
https://doi.org/10.1007/978-3-031-06493-7_11

and, on the other hand, organized groups from many parts of society who see these penetrations as a menace to their productive systems, livelihoods, and health, while also being destructive of their communities and cultures as well as the ecosystems on which they and we all depend.

This chapter addresses some of the underlying causes of these conflicts, by giving voice to actors who are actually involved in developing their own alternatives to the development proposals of the hegemonic forces driving the transformations in their societies. These alternatives emerge from groups whose organizations are shaped by different cosmologies, products of their multiple ethnic origins, and the profound philosophic and epistemological debates of the past half-century that emerged from numerous social movements proposing different strategies for achieving progress, improving well-being, and conserving ecosystems. While many past confrontations among social groups have produced compromises modifying individual development projects, few have created space for the emergence of alternative social and productive structures that respond to the demands for local control of the governance process in order to assure local well-being and responsible environmental management.

The analysis draws on an important emerging literature that proposes a different epistemology and methodology, reflecting the direct participation of a wide diversity of communities around the world and their possibilities for implementing different approaches in order to improve their well-being. In spite of the widely separated regions and traditions from which they come, there are striking commonalities in their reflections on how research should be conducted and how they might collaborate with "outsiders" in their search for alternative ways to advance in their pursuit of an improved style of life and their ability to govern themselves. A notable early contribution from this intellectual and academic current was published by a Maori sociologist (Smith 2021), reacting to the tendency of scholars from the principal academic institutions in New Zealand to make assumptions about local social structures, production possibilities, and the possibilities of and competence for innovations of their "aborigines".

This approach clarifies the difference between dominant concepts of environmental governance and our understanding of the problem, along with its applicability to the work of the communities with which we are collaborating. As generally understood in Western social science literature, environmental governance is an extension of the process of public deliberation and policy formulation, to integrate into the socio-political parameters additional considerations of the impact of society on ecosystems, locally and globally. This relatively new field of political and social action has become poignantly crucial in recent years, as the depths of the environmental crises that we are living in have made their impact increasingly evident. In our work, we have clearly identified the problem of governance with the challenge of assuring that we examine the origins of the problems and the proposed strategies to address the intimately related matter of social justice.

However, in this chapter, we focus on the contrasting conceptions of the functionality of the political process and the possibilities for change. The dominant conception derives from a vision in which the world economy is central, a behemoth comprised of a variety of national and regional units into a single interconnected network of markets that feed a process of capital accumulation. This network of markets is controlled by a small group of powerful economic interests, backed by their national governments within an international institutional framework that reinforces their control over national and international economies. The prevailing model of international politics and environmental governance is firmly grounded in the dynamics of the global marketplace, the private ownership of property and the means of production, creating an increasingly unequal distribution of income, wealth, and power within societies and on a global scale as well as producing a devastating impact on the environment.

In contrast, our research identifies myriad local and regional groups trying to overcome centuries of repeatedly being relocated to ever more inhospitable regions while also being targets of oppression, as a result of an unequal form of integration, transforming them from independent peoples into victims of colonialism and (inter)national capitalist "development". By emphasizing their rejection of market-driven forces that control and distribute resources, they are seeking to design and implement different approaches for decision-making, based on a set of values that generally privilege collective solutions and well-being over individual gains. These approaches emerge from a different and more complex set of objectives, rooted in historical experience, cultural traditions, and intergenerational relationships and responsibilities that situate their choices in a longer time horizon than that typically considered by the dominant methodologies currently guiding environmental governance. Because they attempt to bring to the center of social life politics and ethics, displacing from it the economy, they explicitly reject the primacy of an economic calculus in making fundamental decisions about society, economy, or ecosystem management. As a consequence, their decisions often result in proposals that are at odds with the policy prescriptions offered by the institutions with which they must interact, whether it is for the management of specific natural resources or for addressing problems of political, social, and/or economic dynamics. In the process, they are seeking to isolate themselves from the hegemony of these international forces and epistemologies, forging their own institutions to create spaces of greater autonomy, in political, social, and productive spheres while defending their ways of life and their territory from assimilation into the international economy or its outright seizure/appropriation by international capital.

These communities, as examined in later sections of this chapter, are searching for new ways to strengthen their societies and improve their ability to govern themselves. In many cases, this involves a redefinition of their identities, combining knowledge of their cultural heritage with present-day understandings of the significance of their cultural roots and the history of their struggles against many of the numerous forms of injustice to which they continue to be subjected. These struggles have "never been a blind, spontaneous reflex to objective economic conditions. [Rather, they have] been a conscious struggle of ideas and values all the way" (Thompson 1959, 110). As such, the communities have been able "to hold fast to the vision of collective good".

It is striking that a common feature of solidarity in many of these communities is a growing realization of the importance of this heritage and history, its contribution to their own definitions as peoples, as communities, whose collective identities and belief systems have generated unique forms of organization and social dynamics. These organizations are discovering new ways of integrating their belief systems, their cultures, and their relationships to their environments into cosmologies that lead to creating contrasting models of society, models that directly address the demands for social justice and sustainability while protecting the whole panoply of traits that define a people. While the current uncertainties have encouraged the emergence of different forms of localism, isolationism, and often violent fundamentalism, most communities try to discover in their traditions inspiration and wise and sensible alternatives for their current predicaments (Barkin and Lemus 2014).

While forging these new models of society, communities are actively engaged in a complex process of defining (or redefining) their identities. This search for identity is complex, involving the combination of numerous concentric and competing contexts, coming from national and regional or local cultures, ethnic origins, and environmental features that impact on social structures. The demand for social justice, for example, cannot consent to the idea that profound inequalities are part of the human condition; or that changes in the legal system can legitimate the plunder of community resources or planetary equilibriums. This discussion necessarily leads to a profound distinction between the nature of the social contract on which each society is constructed, posing the question of whether the individual has the right, in the ultimate instance, to assert her (or his) individual interest at the expense of the community's, a right which is generally questioned within the communities with which we are collaborating. For many of them, they are not individuals but singular persons, knots in nets of relations, for whom the community is the first layer of their personal being (Villoro 2003).

Of course, these discourses also define trajectories for social progress. The dominant market-based approach identifies an increase in material production as the leading indicator. Economic growth, as valued in the marketplace and measured by monetary units aggregated into indices of GNP, clearly devalues changes in the status of women, the well-being, or the impact of production on natural resources and ecosystems. In contrast, the version emerging from Latin American community initiatives generally incites broader discussions about lifestyles and community organization, and identifies moderation as part of complex strategies for constructing alternative organizations. Our consultations with the communities to which we refer in this chapter identified five basic principles for this process: autonomy, solidarity, self-sufficiency, productive diversification, and regional sustainable management.

In what follows, we summarize our direct collaboration with communities and alliances of local groups involved in the process of trying to consolidate their own governance structures capable of responding to their visions of an appropriate society consistent with assuring well-being and sustainability. Their struggles seek to consolidate alternative programs to produce the basic goods needed to assure their livelihoods and strengthen their ability for self-governance, while attempting to respect the possibilities and limits of their environments. What is striking about these collaborations is the extent to which the participants are well-informed of the burgeoning discussions of epistemologies that explicitly question the logical structures of dominant governance and development models; many of these seemingly academic debates have become an integral part of the discussions and design of strategic proposals by these local groups to understand and implement programs for local and regional advance.

While most of the detailed fieldwork that we are documenting is based on intensive interactions with communities in the Mexican state of Oaxaca, the materials for this chapter draw on additional contributions produced by people involved in local and regional processes in other parts of the region and with others who are emerging from resistance movements to implement their own proposals for consolidating a material and institutional basis for improving material well-being and assuring their capability for promoting a balanced ecosystem.

2 An Alternative Understanding: A Different Point of Departure

Forging their own solutions is an ambitious endeavor for peoples proposing to overcome discrimination, marginalization, and systematic efforts by colonial powers or by today's power elites to relegate them into ever more isolated corners of their territories. What is remarkable about these efforts is the wealth of proposals with which they are experimenting and the tenacity with which they continue to resist efforts to integrate them into national and international economies as underprivileged individuals in increasingly polarized societies.

Although we concentrated our efforts on collaborating with groups in a limited number of regions in Mexico with high concentrations of ethnic populations, it quickly became clear that resistance movements are proliferating throughout the hemisphere, partly in reaction to State policies promoting local integration into national and international development projects, by permitting outsiders privileged access to natural resources and to construction of infrastructure, in territories traditionally controlled by these peoples. What we found, however, was that there are also positive developments motivating communities throughout the Americas to strengthen their abilities to govern their territories, by better understanding the relationships between themselves and their surroundings while also engaging in deliberate efforts to build alliances among themselves and transnational organizations capable of defending their claims in international arenas.

The need for this process of organization has become increasingly evident as conditions within each country change dramatically with globalization. A concerted effort to accelerate the region's internal integration and connectivity with the global economy as well as to facilitate the access of international enterprises to domestic resources as part of a drive to promote domestic growth is altering the map of Latin America (Bessi and Navarro 2014), impacting first and foremost indigenous communities in the hemisphere. These analysts summarized the problem:

> The reordering of territory has blurred borders in both economic and political terms with projects such as the Mesoamerican Project (previously Plan Puebla-Panama) and the Initiative for Regional Infrastructure Integration of South America, which both entered into force after 2000. Their primary objectives include the construction of transportation and telecommunication networks, as well as energy-generation projects such as hydroelectric dams and wind farms. They also plan to designate national parks, protected areas, Heritage for Humanity sites, cross border conservation areas, transnational parks (also called Parks for Peace), ecological and biological corridors and networks of protected areas... The design of these projects is indeed strategic, and 'progressive' governments are presenting them as a development opportunity. (in Navarro and Bessi 2014)

Ana Ester Ceceña, a Mexican economist, added (in Bessi and Navarro 2014):

> What will happen with IIRSA (Initiative for the Integration of the Regional Infrastructure of South America) is that local governments will be forced to be more disciplined because they will be brought in line with global markets. There are 500 transnational companies that produce half of global gross domestic product; when one looks at IIRSA's design and these companies' projects, they complement one another: The groundwork is being laid for the circulation of communication, merchandise, raw materials and energy... Capital needs a reordering of territory - considering this as a type of historical-social construction - in order to continue reproducing itself, as much in terms of materials as in power relations, of accumulation of capital and profits. The ordering enables access on a large scale to certain types of material from the earth.

Indigenous peoples are increasingly insistent on demanding the recognition and integrity of their territories, many of which are threatened by the grandiose proposals of global capital; their actions are confronting directly these schemes, and changing the maps of the Americas in the process. They have strengthened their resolve to prosecute their historical claims as they become increasingly skilled in achieving the enforcement of the agreement ratified at the International Labor Organization (ILO) to guarantee prior consent of native peoples with territorial claims for outsiders to undertake activities or exploit natural resources in their regions. Accompanying the changing map is a new consciousness of the significant differences in understandings of the most elemental concepts in their exchanges with their interlocutors in the States of which they are a part.

Although a significant discrepancy occurs throughout the Americas, as different social groups and peoples question governmental procedures to charge a single agency with implementing unified policies for the myriad ethnic groups in their countries, an even more serious source of conflict involves the very notion of property and the apparent freedom with which outsiders (government agents) can discuss the possibility of alienating people's claims to land or natural resources. This problem arises because of the profound differences between the historical significance attached to the different concepts of property and territory, which are not easily incorporated into prevailing market-based understandings of the significance of land or property. This is so essential that even the Organization of American States finds itself obliged to take note of its consequence in the context of the demand to draft an American Declaration of the Rights of Indigenous Peoples. This discussion is central to our understanding of the underlying basis of the prosecution of demands for autonomy by native peoples:

> Territorial rights are a central claim for Indigenous Peoples in the world. Those rights are the physical substratum for their ability to survive as peoples, to reproduce their cultures, to maintain and develop their organizations and productive systems... Indigenous Peoples have strengthened their organizations and developed a more organized struggle to reclaim their rights. Central among those demands are the issues related to land, territories and natural resources... these rights are not merely a real estate issue...Rather indigenous land rights encompass a wider and different concept, that relates to the collective right to survival as an organized

> people, with control of their habitat as a necessary condition for the reproduction of their culture, and for their own development, or as Indigenous experts prefer, for carrying ahead "their life plans" and their political and social institutions (Kreimer 2003, 1–2).

Indigenous areas, then, are a complex amalgamation on which the very existence of these peoples depends. This is clearly defined in the Brazilian constitution which gives renewed strength to the ancestral possession as basis for the territorial rights characterized by four significant traits: (1) permanent ancestral possession; (2) areas necessary for their productive activities, including the reproduction of flora and fauna; (3) areas necessary for their cultural reproduction, and for their survival as a collective; and (4) habitat with the physical capacity and shape to allow the full functioning of the mechanisms of authority and self-government of the Indigenous People. These territories are the habitat necessary for their collective life, activities, self-government, and cultural and social reproduction.

Problems arise when the State seeks to exercise its sovereignty or eminent domain, to build infrastructure, to exploit or license the exploitation of natural resources, or any other action or project that may affect indigenous lands and the use of their territory. International law now restricts this possibility, obliging the previous fair and serious consultation with the affected indigenous peoples (Convention 169, ILO, endnote vii). Since indigenous peoples are consolidating their constitutional and legislative demands to codify symbolic and political elements of autonomy and self-government, as elements of internal self-determination, governments are finding themselves treading on new "ground" as they attempt to reconcile global visions of "development" with local efforts to achieve well-being.

Throughout the Americas, governments continue to assume that prices of both landed property and natural resources can be fixed according to market processes and, in the best of circumstances, negotiators of good will can arrive at mutually beneficial agreements for their exploitation, thus assuring their "unlocking" to promote national development by trading them in the global marketplace. In these circumstances, it seems almost incomprehensible to the dominant powers that local groups might object to the terms of these negotiations, refusing to even discuss the possibility of emplacing a forest enterprise, a mine, or a power generating facility in their regions, as it would upset a delicate historical and spiritual balance that they consider threatening to their social structure or cultural integrity, defined in terms of one or more many non-monetary dimensions for which financial compensation is inconceivable.

The nature and scope of this struggle is very old. The old tradition of illegal settlement, which shaped most Latin American cities during the XX century, is today complemented with active movements to redefine urban life. Territorial defense is also a new central theme in the cities. The most spectacular case was Argentina (2001–2002), but from Oaxaca (2006) to Brazil (2014) vibrant movements express the vitality of new social subjects and new forms of social protagonism (Colectivo Situaciones 2002; Mariotti et al. 2007; Zibechi 2008; Giarraca and Teubal 2009).

3 Building the Commons: Local Solutions Are Collective Endeavors

The complex process of differentiating territory from property and clarifying the significance and importance of social ownership and membership as distinct from individual activities encompasses yet another important dimension: the communities generally think of themselves as part of a regional and even a global commons. But unlike the discussions of the concept in much of the academic literature, their understanding of the commons cannot simply be reduced to a collection of "common pool resources" such as air, waters, and other natural resources shared by all that were the focus of the debate set off by Garret Hardin's "tragedy" (1968). Rather, their activities are much more akin to what one of the leading historians of the process describes as the "active movements of human commoning and the worldwide demands to share wealth and safeguard common resources on every continent" (Linebaugh 2008, 280). The organizations that are so engaged are not involved in shaping "an alternative economy, but rather an alternative to the economy" (Esteva 2014, 149). The commons are extended to encompass the social and cultural components of collective life; they are not simply a set of things or resources. Rather, the organizations they are creating bestow great importance on social relations within the community as well as a firm commitment to ensuring the conservation and even the enlargement of the commons. This relationship reflects a collective and enduring transformation of the way in which society conceives and manages itself while also developing the basis for collective and communal management (Bollier and Helfrich 2012).

Protecting, defending, and governing the commons are complex and risky processes. Complex, because doing so encompasses all aspects of social and biological existence. Risky, because it involves challenging the de facto powers and questioning the legitimacy of their "rule of law;" that is, the legal system that is creating and perpetuating a profoundly unjust society, exacerbating social disparities, and accelerating environmental destruction. This dispute about the nature of the State stems from a rejection of the philosophical underpinnings of the hegemonic order, based on the idea of a single "social contract" that presupposes the possibility of applying universal norms, like "social justice", "equality", or even "democracy", impartially to attend to the needs of all social groups. For this reason, it also involves a prima facie repudiation of the legitimacy of national "authorities", who assume their right to transfer community resources—the commons—to others for whatever reason, without regard for the well-being of the people, local decisions, or historical and environmental considerations, as is common practice in mining, forestry, and water management, although it now extends to complex issues of bio- and nano-technology in many nations today. Thus, the efforts to promote solidarity among diverse social groups call for a political approach that requires each to extricate itself from the dominant social and political institutions that are incapable of attending to their particular needs.

However, consolidating the foundations of this society entails much more than undertaking specific activities or establishing appropriate institutions for governance or management. The solidarity society requires personal commitments from each member to assume responsibility for the well-being of others and for limiting individual claims for access to collective resources (Martinez Luna 2010; Robles Hernández and Cardoso Jiménez 2008). To strengthen these foundations, it is essential to begin with a common vision of society as a whole whose point of departure is reversing the historical tendency for the personal enrichment of a few at the expense of the many. As such, they incorporate collective decisions to assure transparency and direct participation in decision-making and universal responsibility for administration or implementation of this dynamic. In historical terms, and specifically in light of practice in today's globalized society, it calls for a redefinition of peoples' relationship with their society, rejecting the notion that one person has the unfettered right to withdraw from, or even oppose, the commonweal after having participated in the process of arriving at a decision.

This point of departure has important implications for the way in which priorities are determined and activities are organized. Perhaps one of the most striking and demanding of these is the need to reverse the hierarchical organization of the workplace. Of course, people should be paid for their work, but they should not have to submit to demeaning and authoritarian social relations to satisfy their basic needs. The existing proletarian organization of society is part of an underlying condition of the helplessness of the workers, unable even to survive without entering the labor force. The alternative under construction here starts with the presumption that all members of society enjoy the legitimate right to a socially determined way of life, independent of their contributions to production or output. Their participation in collective activities becomes rooted in a sense of duty and belonging to the community, but also an obligation that is explicitly enforced by communal authorities. Such an approach eliminates the double alienation of modern labor: from the fruits of work and from the logic of creative activity.

4 Creating the Foundations for Communal Governance: Generating and Managing Surplus

The decision to create autonomous forms of self-government within the framework of the nation-state represents an audacious challenge to the prevailing model of governance and of social and economic justice based on representative democracy and its marriage with the free market. Rooted in the commitment to define and defend their territories, the process involves creating new institutions and processes for the social appropriation of both the natural environment and the productive systems that they have created to assure their ability to maintain and strengthen their community, to provide for their basic needs, and to facilitate exchanges with partners (barter) and in the marketplace. The mechanisms established by the communities for management often involve complex dynamics for mutual consultation among different groups within the communities as well as forms for delegating responsibilities to members on the basis of expertise and social commitment, or for assuring broad political participation and accountability. Thus, it is not only the choice of activities themselves but also the implementation processes that are crucial for the design of the social mechanisms that contribute to the desired outcomes related to equity and sustainability. In the following discussion of individual projects (see the next section), an interesting facet of the analysis is not only the choice of technique but also, and often just as important, the nature of the activities themselves; they speak to a concern for addressing the socially defined basic needs of people in the communities while also creating a balance between the use of natural resources and the restoration, regulation of land use, and conservation of the ecosystems from which they are drawn.

What makes these activities unique is that they are organized by groups who come together on a voluntary basis to ensure their viability and continuity. In many cases, they are trying to regenerate the social fabric eroded by both external and internal forces. While we focus on the collective nature of decision-making, it is just as significant to understand the mechanisms that make possible the consolidation of the community and its ability to advance. During our interactions with the communities in their search for solutions that provide the wherewithal for moving forward we identified a central feature that contributed to this success—one that also explains their ability to consolidate the capacity to implement the collective governance model that is fundamental for society's continuity and its possibility to assure improvements in the lives of its members: the explicit organization of social and productive resources to generate surpluses for "reinvestment" and "redistribution" (Baran 1957).

The centrality of surplus in community management is an often invisible and misunderstood facet of the administrative process. Much of the literature describes rural communities in general and indigenous groups in particular as living at the margins of subsistence, as the poverty in material means limits their ability to advance and reduces the scope for broadening the range of activities they can undertake. In contrast, our dealings with communities throughout the Americas reveal their ability and commitment to produce this surplus and manage it collectively, using it to reward members who have made important contributions in producing it and channeling the rest for collective purposes.

By focusing on the processes of producing and managing surplus within the limits for satisfying socially defined needs and the possibilities of their ecosystems, this collective management structure of the diverse local projects has proved effective in constructing a framework for environmental justice that is proving so elusive in the larger societies of which they are a part. Unlike those other parts of society closely tied to the global market economy, these communities have created possibilities for organizing themselves to ensure that their members need not suffer from extreme poverty and unemployment. As a result, they are generating a productive potential far greater than might be appreciated by a simple accounting of the financial resources that they have at their command. Some of this potential is well documented in the literature, as is the case of the "voluntary" labor that is expected from all members for collective tasks involving building and maintaining infrastructure or conserving ecosystems. The social mechanisms for assigning and rotating administrative and political positions so important for governance is another way in which resources that are often invisible in the market economy or formal accounting calculus are generated in these communal organizations. But just as important, the commitment to universal inclusion or participation also creates a corresponding responsibility from the members to contribute to collective tasks—assuring that most individuals will be involved in a multiplicity of activities for their own benefit and that of the community.

Surplus has existed in human organizations from time immemorial. Even when there were no formal institutions for exchange and accumulation, the construction of large and small projects to channel water or create monuments is testimony to the ability of societies to advance beyond their immediate needs, building projects to increase productive capabilities or the grandeur of their "leaders". What distinguishes the myriad communities guided by cosmologies removed from those based on material gain and individual benefit at the expense of the whole is their ability to promote a broad participation for advancing the general welfare. Most recently, these societies have improved their possibilities for implementing new projects, taking advantage of advances in science and technology while also critically incorporating knowledge and contributions from the past, generating opportunities for increased or more efficient production as well as more effective means for improving their well-being and ability to protect their ecosystems. By examining the availability and mobilization of surplus, the communities are better equipped to consider how best to implement their long-term visions. What is striking about the individual experiences related in this chapter is the clear understanding by many of the participants engaged in local problem-solving and the ways in which leaders have implemented activities to achieve change.

5 Communal Approaches to Environmental Justice

Communities across the Americas are involved in designing and implementing local solutions that contribute to their broad struggle for environmental justice. While a great deal of energy must be devoted to protecting themselves from encroachment by forces attempting to control their natural resources and subject them to the various disciplines of markets and political systems, it is remarkable that they continue to mobilize locally and nationally while associating internationally with other communities and non-governmental organizations to consolidate new lines and technologies of production and experiment with ways to improve existing activities.

These actions are the product of the complex interaction of dynamic forces within the communities and reactions to outside pressures. They are part of a search for a unique identity that has become increasingly important as these peoples assert their legally binding rights to self-determination as defined by their varied histories and their understanding of the privileges accorded to them under the ILO Convention 169 and similar agreements promulgated in other international bodies, and the ongoing efforts in the Organization of America States (2003) to draft a similar commitment (endnote ix). In Mexico, as elsewhere, this process has a long history, which was codified in the Constitution of 1917, as indigenous communities were recognized and granted collective rights by the agrarian reform.

During the last half of the XX century, Mexican communities waged an unrelenting and difficult battle to assert their rights to control the lands over which they were able to retain or regain control after the Revolution. They were particularly effective in wresting control of their communal forests from private firms that had been exploiting them for decades (Barkin and Fuente 2013; Stevens et al. 2014). Today, there are a variety of management plans in effect, testimony to skills that the communities have acquired as they attempt to reconcile pressures for ensuring conservation with the need to create jobs and generate incomes. The movement to reassert indigenous identities in Mexico was further strengthened in the aftermath of the 1994 uprising in Chiapas by the Zapatista Army of National Liberation (EZLN) (Muñoz Ramírez 2008).

Since then the activity and visibility of indigenous peoples throughout Mexico has increased along with a gradual recognition of their importance in the population, because of and in spite of the growing intensity of repressive actions by the State and other actors, including private corporations given concessions in these territories and organized groups in various parts of the society. While a recounting of the initiatives being implemented in these communities would be too lengthy for inclusion here, suffice it to say that the discussion of many of them within the framework of the National Indigenous Congress, and the increased circulation of information and meetings among members contribute to strengthen the resolve and ability of members to carry their projects forward.

In connection with their efforts to gain recognition and elaborate local management strategies, control of water resources has been particularly contentious as communities try to assert their rights to adequate supplies and protect their sources. These movements are now inextricably combined with others in opposition to large-scale construction projects for dams designed to harness waters for electricity generation or for long-distance transfer between water basins to supply urban areas where aging infrastructure and excessive growth in consumption are causing shortages due to lack of administrative and technical capabilities of dominant bureaucracies. As a result, many communities that have historically been able to satisfy their own needs and even share surpluses with neighboring communities are now finding themselves involved in coalitions with others defending their water sources, along with ecologists who are generally arguing that the engineering and public works approaches of the public sector are inappropriate and simply postponing the day of reckoning with regard to the need for a more ecologically informed approach to water management.

An interesting outcome is the combination of traditional and leading-edge technologies applied to protect their natural sources—the streams and springs on which they depend. This combination of technologies with direct community involvement in water management contrasts sharply with the national water authorities' approach that eschews local diversity, preferring a homogenous administrative model conducive to centralized management and engineering solutions. In response to the great differences in local conditions, there are many examples of water-saving technologies being implemented by the communities, such as installing composting toilets and separating grey from black water flows to allow for low cost and passive biological processing conducive to restorative environmental practices. The project "Water Forever" transformed a million hectares of barren plateau and steep slopes recreating underground aquifers and structures found in some of the oldest irrigation projects in the Western Hemisphere from the eleventh century. This project, which began in the 1980s, is noteworthy because it combines community-managed agro-ecological and agro-industrial activities and enterprises belonging to the participants, creating jobs and products that prove attractive to consumers for their social, ecological, and nutritional qualities (Hernández Garciadiego and Herrerías Guerra 2008). In Bolivia, the experience of the "Water War" of 2000 in Cochabamba is still vivid in people's memories as local water committees continue to organize actively while resisting the State's effort to manage the commons (Fogelberg 2013; Dwinell and Olivera 2014).

These community-based management proposals embrace important parts of their members' collective existence, but cannot provide for all of the needs of the community. Having adequate water supplies and sustainable models for forest management can build stronger and more resilient communities. Unfortunately, the recent pressures on national governments to increase energy production from renewable sources are heightening the conflicts with indigenous communities threatened with being flooded out of their territories. In Mexico, the refusal of the government to permit indigenous communities to undertake their own micro-hydroelectric power projects is clear evidence of the fear of the degree of independence that such activities would promote.

Despite these obstacles and conflicts in the power and water sectors, numerous communities are undertaking productive activities to supply basic needs and to create goods that can be traded for other products. Ongoing efforts are oriented to identifying new activities that make use of available renewable resources to produce goods that might be advantageously exchanged with others to provide for these basic needs. The objective of this approach is to induce social dynamics that bring the producers together into stronger organizations that become, in turn, parts of their communities.

As part of this effort, many groups accompany communities in introducing complementary activities and assisting them in modifying technologies or introducing new ones that would strengthen their organizational capabilities to contribute to the collective well-being. The objective of these undertakings is to contribute to community efforts to strengthen their own capacity to govern themselves. La Vía Campesina, for example, has a presence in 73 countries representing more than 200 million members. Its purpose is to promote food production by using agro-ecological techniques to move groups of producers toward greater self-sufficiency promoting food self-sufficiency by combining traditional farming practices and knowledge with the most innovative agroecological practices, creating a unique paradigm disseminated in an ongoing educational process in peasant-to-peasant workshops (Val et al. 2019; Rosset et al. 2019). Other social groups are also actively engaged in activities that promote social, political, and productive changes to contribute to improving their own lives as well as those of others while attempting to conserve and enhance environmental quality or sustainability. In Mexico, the local Caracoles in Chiapas contribute to this objective, directly improving the lives of many people while also portraying a model of social organization and change that continues to have a powerful effect on other communities as well as in other countries. Such activities have achieved a high level of self-sufficiency in food, health, and education; thus, they have improved the well-being, contributed to diversifying the economy, and increased productivity in a region where perhaps as many as 500,000 people participate (Baronnet et al. 2011; Vergara-Camus 2014; Mora 2015).

In South America, Andean communities are similarly involved in promoting collective strategies known as "buen vivir" (Huanacuni 2010; Beling et al. 2021). Throughout the Americas, groups of communities are involved in mobilizations to defend their territories, cultures, and societies: groups like "Idle no More" in Canada, the Haudenosaunee (Iroquois) Confederacy in eastern North America, the Landless Workers Movement (MST) in Brazil, the Mapuches in Chile, and numerous others throughout the region, as well as the National Indigenous Congress, the Network of Environmentally Affected Peoples, and the Movement Against Mining in Mexico. Similarly, there is a coalition of indigenous peoples in the Americas and a series of international non-governmental organizations that are promoting strategies for better resource use but most of the mobilizations help to defend groups against others trying to take control of their resources or organizing to forestall activities that might contaminate their lands or their waters (Vergara-Camus 2014).

Accompanying these actions of resistance, many communities are involved in other constructive activities, promoting collaboration with university and civil society researchers who are contributing to explaining the value of the work, while contributing to diversifying economies and improving production in sustainable ways (Toledo et al. 2013; Toledo and Ortiz Espejel 2014). In a different approach, scholar-activists are working with producers in diverse regions to protect and enhance the growth of traditional products (Illsley Granich et al. 2007). In Peru and more recently Bolivia, the well-established technical promotion and development organization Pratec deploys effective approaches for community-based learning, improving production in the multiple ecologies of the Andean world, focusing on potatoes but carefully balancing its work to support broad-based, diversified progress (Gonzales 2014). Ecotourism is yet another, more controversial activity, because it involves an explicit opening of the community to outsiders who are frequently unable to comprehend the magnitude of the cultural and economic chasm that separates them from their hosts (Barkin 2002; Fuente and Ramos Morales 2013).

Elsewhere, indigenous peoples, peasants and industrial workers are all exploring new routes to reorganize their workplaces and contribute to improving living standards for themselves and their communities. New production systems are being invented as workers occupy closed factories, continuing operations by changing management and incentive systems (Lavaca 2007; Ruggeri 2013; Ness and Azzellini 2011; Piñieiro 2013). In many cases, the initiatives not only placed the direct producers in control of the enterprises but also often created opportunities including the community in decisions and incorporating the impact on the environment into the new decision-making calculus.

6 The Prospects for Alternative Strategies for Environmental Justice

While these initiatives are changing the map of the Americas (Navarro and Bessi 2014), many other developments are threatening to erode the possibilities for improving peoples' lives and taking better care of the environment. Throughout the hemisphere, much environmental governance involves attempts to minimize the deleterious social and ecological impacts of harmful activities that are the foundation of national and international development. Industrial work is intensifying and ever more alienating and labor has fewer protections. Natural resource concessions are opening up vast new territories to exploration and production, with terrible environmental impacts. The privatization of public services and the deterioration in the quality of those remaining in the public sector are a palpable threat to peoples in every country.

Even as indigenous communities are asserting their new found rights to proceed with forestry and water management activities, governments are encouraging large-scale initiatives by transnational corporations that threaten to upset the delicate balance of productive activities on which the communities depend for their livelihoods and for ecosystem balance. These projects pose fundamental questions about the ability of the communities to defend their territories, including their substantial cultural, social, and productive heritage that entrenches them in their ecosystems. The conflicts continue to this day, posing apparently irresolvable differences and oft-times violent encounters, as mines, ecotourism, and other projects threaten the very existence of the communities (and with the recent reforms, fracking and other forms of resource extraction). The communities generally reject the assumption that the sacrifices that this destruction entails can be compensated by monetary offers that would only force them onto a path of institutionalized marginalization as isolated individuals, a life of limited opportunities without the social support systems and safety nets that their communities offer.

The ongoing initiatives to strengthen or generate "niches of sustainability" by peasant and indigenous communities throughout the Americas are heartening and important. While the momentum in the global marketplace clearly threatens social groups and environments everywhere, the continuing successful efforts of peasants and indigenous peoples to implement their own strategies for social and productive change that deliberately incorporate the environment in the process offer a window on the possibilities for making environmental justice a reality for increasing segments of the population. Thus, the implementation of local solutions that create regions for autonomous action will become even more significant and effective as the spaces dominated by the global market continue to suffer from deteriorating environments and heightened conflict.

References

Adrián E. Beling, Ana Patricia, Cubillo-Guevara Julien, Vanhulst Antonio Luis, and Hidalgo-Capitán. *Latin American Perspectives*, 48(3) (2021): 17–34. https://doi.org/10.1177/0094582X211009242.

Baran, Paul A. *The Political Economy of Growth*. New York: Monthly Review, 1957.

Barkin, David. "Indigenous Ecotourism in Mexico: An opportunity under construction," in Deborah McLaren (ed.), *Rethinking Tourism and Ecotravel*. Westport, CT: Kumarian Press, 2002.

———. "Communities Constructing Their Own Alternatives in the Face of Crisis," *Mountain Research and Development*, 32(S1) (March 2012): S12–S22.

Barkin, David, and Blanca Lemus. "Rethinking the Social and Solidarity Society in Light of Community Practice," *Sustainability*, 6 (September 2014): 6432–6445.

Barkin, David, and Mario Fuente. "Community Forest Management: Can the Green Economy Contribute to Environmental Justice?" *Natural Resources Forum*, 37 (August 2013): 200–210.

Baronnet, Bruno, Mariana Mora Bayo, and Richard Stahler-Sholk. *Luchas muy otras: Zapatismo y autonomía en las comunidades indígenas de Chiapas*. Mexico: UAMX-CIESASUNACH, 2011.

Bessi, Renata, and Santiago F. Navarro. "The Changing Map of Latin America," July 2, 2014. http://www.truth-out.org/news/item/25227-the-changing-map-of-latin-america.

Bollier, David, and Silke Helfrich. *The Wealth of the Commons*. Amherst, MA: The Levellers Press, 2012.

Bonfil Batalla, Guillermo. *Mexico Profundo: Reclaiming a Civilization*. Austin, TX: University of Texas Press, 1987 [1996].

Colectivo Situaciones. *19 y 20: Apuntes para el nuevo protagonismo social*. Buenos Aires: Colectivo Situaciones/Ediciones mano a mano, 2002.

Dwinell, Alexander, and Marcela Olivera. "The Water Is Ours Damn It! Water Commoning in Bolivia," *Community Development Journal*, 49(S1) (2014): i44–i52.

Esteva, Gustavo. "Commoning in the New Society," *Community Development Journal*, 49 (Suppl. 1) (January 2014): i144–i159.

Fogelberg, Kate. "From Adopt-A-Project to Permanent Services: The Evolution of Water for People's Approach to Rural Water Supply in Bolivia," *Water Alternatives*, 6(2) (2013): Article 6. http://www.water-alternatives.org/index.php/alldoc/articles/vol6/v6issue3/222-a6-33?path=articles/vol6/v6issue3.

Fuente, Mario, and Fernando Ramos Morales. "El ecoturismo comunitario en la Sierra JuárezOaxaca, México: entre el patrimonio y la mercancía," *Otra Economía. Revista Latinoamericana de Economía Social y Solidaria*, 7(12) (2013): 66–79.

Giarraca, Norma, and Miguel Teubal (Coords.). *El campo argentino en la encrucijada: Estrategias y resistencias sociales, ecos en la ciudad*. Buenos Aires: Alianza Editorial, 2005.

Gonzales, Tirso. "Kawsay (Buen Vivir) y afirmación cultural: Pratec-Naca, un paradigma alternativo en los Andes," in Boris Marañon (ed.), *El Buen Vivir y descolonialidad: Critica al desarrollo y la racionalidad instrumentals*. México: UNAM-Instituto de Investigaciones Económicas, 2014.

Hardin, Garret. "The Tragedy of the Commons," *Science*, 162 (December 1968): 1243–1248.

Hernández Garciadiego, Raúl, and Gisela Herrerías Guerra. "El programa Agua para Siempre: 25 años de experiencia en la obtención de agua mediante la regeneración de cuencas," in Luisa Paré, Dawn Robinson and Marco González Ortiz (eds.), *Gestión de cuencas y servicios ambientales. Perspectivas comunitarias y ciudadanas*. México: Instituto Nacional de Ecología, 2008.

Huanacuni, Fernando. *Buen Vivir / Vivir Bien. Filosofía, políticas, estrategias y experiencias*. Lima: Coordinadora Andina de Organizaciones Indígenas—CAOI, 2011.

Illich, Ivan. *Tools for Conviviality*. New York: Harper & Row, 1973. https://monoskop.org/images/3/3e/Illich_Ivan_Tools_for_Conviviality_1975.pdf.

Illsley Granich, Catarina, et al. "El Grupo de Estudios Ambientales, AC: Entre la acción social y la consolidación institucional," in Anthony J. Bebbington (ed.), *Investigación y cambio social. Desafíos para las ONG en Centroamérica y México*. Guatemala: School of Environment and Development et al., p. 25 (2007): 175–208.

Kreimer, Osvaldo. "Indigenous Peoples' Rights to Land, Territories, and Natural Resources: A Technical Meeting of the OAS Working Group," *Human Rights Brief*, 10(2) (2003): 13–17.

Lavaca Collective. *Sin Patrón: Stories from Argentina's Worker-Run Factories*. Chicago, IL: Haymarket Books, 2007.

Linebaugh, Peter. *Magna Carta Manifesto: Liberties and Commons for All*. Berkeley: University of California Press, 2008.

Mariotti, Daniela, María Comelli, María Inés Petz, Juan Wahren, Norma Giarracca, and Miguel Teubal. *Tiempos de rebelión: "Que se vayan todos," Calles y plazas en la Argentina: 2001–2002*. Buenos Aires: GEMSAL, 2007.

Martínez Luna, Jaime. *Eso que llaman comunalidad*. México: Conaculta, 2010.

Muñoz Ramírez, Gloria. *The Fire and the Word: A History of the Zapatista Movement*. San Francisco, CA: City Lights Press, 2008.

Navarro, Santiago F., and Renata Bessi. "Across Latin America, a Struggle for Communal Land and Indigenous Autonomy," July 2, 2014. http://www.truth-out.org/news/item/24981across-latin-america-a-struggle-for-communal-land-and-indigenous-autonomyn.

Ness, Immanuel, and Dario Azzellini (eds.). *Ours to Master and to Own: Worker Control: From the Commune to the Present*. Chicago, IL: Haymarket Books, 2011.

Organization of American States, Committee on Juridical and Political Affairs (OAS-CJPA). Working Group to Prepare the Draft American Declaration on the Rights of Indigenous Peoples (OEA/Ser.K/XVI, GT/DADIN/doc.113/03 rev. 1, 20 February 2003). Report of the Rapporteur, "Traditional Forms of Ownership and Cultural Survival, Right to Land and Territories." 2013. http://www.oas.org/council/CAJP/Indigenous%20documents.asp#2003. Accessed 10 Sept 2014.

Piñeiro Harnecker, Camila. *Cooperatives and Socialism: A View from Cuba*. London: Palgrave Macmillan, 2013.

Robles Hernández, Sofía, and Rafael Cardoso Jiménez. *Floriberto Díaz. Escrito, "Comunalidad, energía viva del pensamiento mixe."* México: UNAM, Programa Universitario México Nación Multicultural (La Pluralidad Cultural en México; 14), 2008.

Rosset, Peter, Valentín Val, Lia Pinheiro Barbosa, and Nils McCune. Agroecology and La Via Campesina II. Peasant Agroecology Schools and the Formation of a Sociohistorical and Political Subject, *Agroecology and Sustainable Food Systems*, 43(7–8) (2019): 895–914. https://doi.org/10.1080/21683565.2019.1617222.

Ruggeri, Andres. *Cuadernos para la Autogestión*. Buenos Aires, Argentina: Chilavert, 2013.

Smith, Linda Tuhiwai. *Decolonizing Methodologies: Research and Indigenous Peoples*. 3rd Edition. London: Zed Books, 2021.

Stevens, Caleb, Robert Winterbottom, Katie Reytar, and Jenny Springer. *Securing Rights, Combating Climate Change: How Strengthening Community Forest Rights Mitigates Climate Change*. Washington, DC: World Resources Institute, 2014.

Thompson, Edward P. "Commitment in Politics," in Cal Winslow (ed.), *E. P. Thompson and the Making of the New Left*. London, UK: Lawrence & Wishart, [1959] 2014. Available online at: https://www.marxists.org/archive/thompson-ep/1959/commitment.htm

Toledo, Víctor Manuel, and Benjamín Ortiz Espejel. *México, Regiones que Caminan Hacia la Sustentabilidad: Una geopolítica de las resistencias bioculturales*. Puebla: Universidad Iberoamericana, Campus Puebla, Conacyt, 2014.

Toledo, Víctor Manuel, David Garrido, and Nicolás Barrera Bassols. "Conflictos Socio-Ambientales, Resistencias Ciudadanas y Violencia Neo-Liberal en México," *Ecología Política*, 46 (2013): n115–n124.

Vergara-Camus, Leandro. *Land and Freedom. The MST, the Zapatistas and Peasant Alternatives to Neoliberalism*. London: Zed Books, 2014.

Villoro, Luis. *Creer, Saber, Conocer*. México: Siglo XXI, 2003.

Zibechi, Raúl. *Autonomías y emancipaciones: América Latina en movimiento*. México: Bajo Tierra Ediciones, 2008.

Linking Caribbean Development Options to Green and Blue Economic Growth: Key Notions and Policy Implications

Nikolaos Karagiannis and Debbie A. Mohammed

1 Introduction: The Caribbean Development Context

Historically, the Caribbean countries developed under the tutelage of different European empires, and more recently under North American dominance. As a result, the Caribbean came to be a classic area of plantation society as there is a certain unity to the region's development and in its patterns of historical evolution. Furthermore, the ***plantation economy*** models emphasized the historical continuity of Caribbean dependence from the slave plantation to modification following emancipation, to further modification in the post-colonial era. Political independence established national sovereignty in older and newer nations of the

N. Karagiannis (✉)
Winston-Salem State University, Winston-Salem, NC, USA
e-mail: karagiannisni@wssu.edu

D. A. Mohammed
International and Global Business, Centennial College,
Toronto, ON, Canada

N. Karagiannis and J. E. King (eds.), *Visions and Strategies for a Sustainable Economy*, Global Institute for Sustainable Prosperity,
https://doi.org/10.1007/978-3-031-06493-7_12

Commonwealth Caribbean, when both groups were integrated into the international system. Consequently, the political process of national independence created states, societies, and nations that had evolved as integral parts of the global system. In the case of the newer nations, independence coincided with the beginning of a new era that was defined by the decline of the old great powers and the rise of two superpowers with antagonistic ideologies and strategic interests: the United States of America and the Soviet Union.

Recognizing the disadvantages of small-scale economies and seeking to re-examine their position on the globe, Caribbean states made efforts to cooperate and push for deeper regional integration. Accordingly, Caribbean leaders met in Chaguaramas, Trinidad in 1972, and created the Caribbean Community and Common Market (CARICOM). CARICOM came into official existence in 1973 with a protocol based on the Georgetown Accord. CARICOM's main goals are economic integration through a common market and common trade policies, functional cooperation (pooling resources and sharing services), and coordination of foreign policies. Then, in 1989, Caribbean leaders in Grenada decided to work towards the establishment of the CARICOM Single Market and Economy (CSME). This thirteen-member organization came into effect in 2006, and its completion took place between 2008 and 2015.

CSME was designed as a single market to facilitate the pooling of the region's human, financial, and natural resources to build the economic capacity required to effectively respond to globalization and the mega trading blocs. Undoubtedly, Caribbean countries continue to struggle to maintain a firm foothold in today's world. In general, there are several similar characteristics and challenges shared among the different economies of the region such as high levels of imports and constraints to exporting, limited natural resources, and underdevelopment. Despite their economic similarities and excessive reliance on foreign investment and aid, each nation's economy exhibits differences too, as some Caribbean economies are more sector-diverse than others. For all, the small scale of each country's domestic economy, underdevelopment, unemployment and underemployment, and pressing social problems in the Caribbean are major challenges to confront in the future.

Since the 2000s, orthodox development policy proposals by international organizations (e.g., US International Trade Commission, UNEP, IMF, IDB, and World Bank) and local economists have been offered as "modern" economic solutions to the region's complex problems.

However, these orthodox policy recommendations—which by and large focus on total factor productivity, resource use efficiency, market-conforming policy suggestions, and more efficient governance—have not resulted in significant socioeconomic development, local industrial growth, endogenous competency, and competitiveness. Furthermore, climate change can be expected to have negative effects on the tourism and hospitality industry which continues to be a major source of foreign exchange for many of these countries. According to the *Caribbean Planning for Adaptation to Global Climate Change Project* (CPAGCCP 2020),[1] virtually all the Caribbean territories are highly vulnerable to climate changes and can expect to experience a "linear increase in the number of storms and hurricanes, ... loss of land from a rising sea level, ... increased susceptibility of coastal infrastructure, ... [and] negative impacts in the tourism sector".

For the Caribbean territories, the climate change challenges are even more severe than they are for most other places around the globe because they have an impact on the entire coastal and terrestrial ecosystems. The Barbados Plan of Action adopted in 1994, recognized that small island developing states (SIDS) exhibited "peculiar characteristics and vulnerabilities" which exacerbate their development challenges. According to *Responding to Climate Change in Small Island Developing States* (UNDP 2010), "small geographical area, isolation and exposure" is further impacted by "sea-level rise, altered rainfall patterns, and storm-surges" which "threaten to reverse progress made towards the Millennium Development Goals now and in the future". The already poor state of the Caribbean marine environment restricts the ability of habitats like seagrass meadows and coral reefs to recover from the effects of severe storms. Poor water quality and over-fishing, for example, promote the overgrowth of algae, preventing recovery. With repeated hurricanes occurring over time periods that are insufficient for recovery to occur, this will only get worse. The devastation of coastal environments, particularly seagrass meadows, can also result in long-term losses of the benefits that humans receive from them, such as fisheries support or coastal protection. Damage to these ecosystem services, consequently, impacts human well-being in the Caribbean because people can no longer rely on them for their livelihood and food supply.

[1] https://www.cpacc.org/the-project/.

Within such a context, the primary objective of this chapter is to extract the relevant aspects of the East Asian successes and embed them in an appropriate framework that emphasizes "green" (which creates a sustainable pathway that protects ecological assets) and "blue" (which encompasses economic activities related to seas, oceans, and coasts) growth strategies for Caribbean economies. The chapter comprises three main parts. Section 2 provides an overview of past development strategies in the Caribbean and places special emphasis on present "green" and "blue" economic growth initiatives at a regional and national levels. Section 3 revisits the developmental state argument and discusses this approach as a pragmatic alternative towards addressing self-sustained growth challenges of Caribbean nations. Taking such multifaceted challenges into account, Sect. 4 offers developmental state-based policy areas which are deemed necessary to boosting Caribbean industries' "green" and "blue" growth potential. Brief conclusions end the chapter.

2 Caribbean Development Strategies: Past and Present

During the last five decades or so, the region has tried different development models, ranging from industrialization based on imported inputs and technology, to models of structural change that prioritized tourism and other services, to open regionalism blending regional integration with efforts to boost the export competitiveness of tradable goods and services. However, the main Caribbean trajectory has relied on metropolitan initiatives in investment, technology and marketing, and on continued metropolitan ownership and control of the region's main means of production and domestic demand. Evidently, the political independence of Caribbean territories has not been accompanied by any significant advancement of their national economies: unbalanced economic structures; international specialization based on unequal exchange; excessive foreign financial and technological penetration; chronic current account deficits and a dependent monetary system, are among the outstanding features of the Caribbean "blocked development" or "underdevelopment". Thus, Caribbean economies have been seriously challenged by a lack of local industrial capacity and export competitiveness, which are among the main factors needed for the achievement of sustained growth and endogenous development. Foreign decisions largely determine the growth prospects of Caribbean economies while most local resources, natural, human, and technological, remain basically underdeveloped.

From the 1990s and onwards, Caribbean development efforts have been characterized by what might be called the consolidation of neoliberal globalization, tempered only by the realization that more policy attention had to be paid to human resource development if the new technological imperatives of a globalizing economy were not to pass the region by. Development was seen as a market-driven, private sector-led process. The role of the state should be to meet the demands for "good governance" based on efficiency considerations and imposed by the international financial institutions, thereby fashioned to submissively serve the logic of deregulated competitive markets and integrated global production, led and directed by powerful transnational corporations.

Indeed, under the influence of the neoliberal narrative, the belief that emerged throughout the developing world by the 1990s, including the Caribbean, was that national economic growth could be significantly enhanced via links with major international economic actors, such as multinational corporations, by encouraging foreign direct investment and privatization projects. In this undertaking, the role of the state was essential in facilitating the process for greater economic integration and for securing the emergence of a socioeconomic order in which the market itself functioned as a form of social regulation—the ultimate goal of the neoliberal vision. However, since the late 1990s, the Caribbean has embraced elements of a "green" and "blue" economy through regional initiatives focusing on such crucial issues as renewable energy, climate change, sustainable tourism, ocean ecosystem services, and biodiversity.

At the regional CARICOM level, notable initiatives include:

- The Caribbean Alliance for Sustainable Tourism (CAST) was established by members of the Caribbean Hotel Association in 1997.
- The Caribbean Renewable Energy Development Programme (CREDP) in 1998, and the Caribbean Sustainable Energy Roadmap and Strategy (C-SERMS).
- CARICOM's regional framework for climate resilient development—implemented by the Caribbean Community Centre for Climate Change (CCCCC).
- Caribbean Natural Resources Institute (CANARI) Sustainable Rural Livelihoods Programme was established in 2011. Further regional dialogue was facilitated by CANARI during 2009–2012 to assist

with the formation of the Caribbean's own unique development pathway.

- UNEP's Green Economy Initiative—launched in late 2008. Subsequently, the Caribbean Green Economy Action Learning Group was established in 2010.
- The Green Economy Caribbean Political Advisory Group (GEPAG) was formed in 2014 to undertake research, outreach, and capacity building in the public and private sectors to support different steps of a green economy transition.

At the national level, various initiatives and policies have been developed in several Caribbean countries as leaders and decision-makers seek to support sustainable development, renewable energy, and climate resilience.

- The Bahamas—National Development Plan: *Vision 2040* in 2016
- Jamaica—National Development Plan: *Vision 2030* in 2018
- Guyana—Low-Carbon Development Strategy in 2009
- Dominica—Organic Agriculture and Development of Organic Production (a low carbon climate resilient development strategy) in 2007
- Trinidad & Tobago—Valuation of Ecosystem Services into National Accounting in 2015
- Grenada—Alternative Growth and Poverty Reduction Strategy in 2013
- Barbados—National Green Economy Scoping Study, with the support of the University of the West Indies and UNEP's Green Economy Initiative in 2011

A "green" and "blue" growth approach is expected to address major problems related to the sustainable development with its threefold goal of social, economic, and environmental welfare of Caribbean nations. However, lack of thoroughness and deep understanding of modern planning requirements, partisan politics and a lack of political consensus, declining human resource base and knowledge brain-drain, outdated or inadequate regulatory frameworks, and a lack of institutional capacity and strength, all impede the transition to a green economic pathway. These impediments can be addressed by a Caribbean-type developmental state, which would account for longer-term potential social and economic gains and engineer nationally inspired solutions. As the notion of the developmental role for the state is rather uncommon to the general economic and political culture in the Caribbean, this pragmatic alternative option is carefully deliberated in the subsequent section.

3 The Caribbean Developmental State: A Pragmatic Alternative

The development experience of Japan, South Korea, Taiwan, Hong Kong, Singapore, Malaysia, and, more recently, China suggests that state intervention can accelerate the process of development and industrialization. Hong Kong and Singapore, in particular, are both small states, so, low population size seems not to be an insurmountable obstacle to the success of the developmental state model. The economies of East Asia and China have achieved remarkable growth rates by adopting Western technologies and business systems, while still emphasizing loyalty to society, family, and the nation. In the early stages of these transformations, the state works with the private sector, coordinating efforts in key sectors, resulting in a market economy that delivers the objectives of a national strategic plan. The successful developmental states avoided the mistake of confusing ownership with the need to build technological infrastructure and invest in human capital formation and research. This is important as government ownership and control of industrial production (as practised in the former Soviet Union) usually ensured that ideological or political rather than industrial factors were paramount, which led to economic failures. The successful developmental states used planned market augmentation as their most important instrument of industrial policy and understood that firms must eventually be exposed to international competitive pressures in order to ensure that they continue to drive towards greater efficiency and competitiveness.

Most would agree that government interventions are necessary when the market alone cannot deliver particular socioeconomic objectives. Recent attention has been increasingly focused on the effectiveness of policy interventions, which raises the crucial issue of the capacity of government bureaucracies. As Evans and Heller (2018: 2) note: "Capable, internally coherent, state bureaucracies were crucial to providing predictability". Effective policies "not only require credible commitments but crucially depend on the administrative, technical, and political capacity and capability of policy-makers" (Ahrens 1997: 115). Furthermore, "(d)ifferent kinds of state structures create different capacities for action. Structures define the range of roles that the state is capable of playing. Outcomes depend both on whether the roles fit the context and on how well they are executed" (Evans 1995: 11).

The ability of developmental states to undertake useful strategic interventions was based on a strong state structure and administrative capacity, with highly educated, powerful civil servants. Such states "... concentrate considerable power, authority, autonomy and competence in the central political and bureaucratic institutions of the state, notably their [determined developmental elites, or] economic bureaucracies" (Leftwich 1995: 420). The bureaucrats typically had sufficient scope to take their own initiatives and operate decisively (Amsden 1989; Wade 1990; Onis 1991). These relatively small, tightly organized bureaucratic structures had most of the Weberian characteristics of highly selective, meritocratic recruitment patterns and long-term career rewards, which enhanced the solidarity and corporate identity of the bureaucratic elites. In the best examples, such as Singapore, meritocratic systems attracted the best managerial talent available to the ranks of the bureaucratic elites and ensured a high degree of unity, capability, and professionalism. This institutional strength helped the bureaucracies to resist any short-term political misdirection and gave them authority, although their relative autonomy was tempered by ultimate accountability (Onis 1991: 114, 124). In a Caribbean environment, however, with a weak and politicized civil service and poor cooperation, selective support for strategic sectors could easily degenerate into an instrument of rent-seeking and short-term gains by vested interests.

Both the relatively high level of bureaucratic autonomy and the close public–private cooperation are important features of the East Asian developmental state model. It is important to note, however, that the East Asian developmental state involved a significant element of compulsion exercised by the bureaucrats in securing public–private partnerships that would be hard to justify by the standards of pluralistic democracy. The extraordinary degrees of control exercised by the state over the financial system and the dependence of individual conglomerates on bank finance were instrumental in eliciting compliance with the requirements of the strategic industrial policy. This resulted in the corporatist nature of most developmental states. In the case of Japan, although there were conflicts among various interest groups within its political system, and competition among firms within its economy, "extensive economic and political agreements have been as important to Japan's economic dynamism as competition" (Nester 1991: 57). This approach did serve Japan well as it built its industrial base during the post-war years. This also reinforces the earlier suggestion that many solutions are specific to particular circumstances and may themselves become problems when the circumstances change.

Most of the developmental states possessed considerable leverage over private industries in terms of securing compliance with national strategic goals, while planning was typically based on consultation and serious efforts to reach agreement on the form of intervention. Economic coordination and development in these states were usually managed by institutions with significant technocratic expertise, located at the interface between state and industry. These were the "pilot agencies" of strategic economic direction (Johnson 1981: 9–10, 1982: 26 quoted in Leftwich 1995). Strategic industrial policy is an important component of the developmental state model, but the successful exemplars did not make the mistake of micro-managing and getting involved with operational detail. Well-educated, competent, efficient, and morally committed planners took a long-term view of the future indigenous development of their countries, and their incentive structures were continually revised in light of planning objectives.

"At the broadest level, East Asian governments managed to generate a sense of commitment to [...] national development", which gained widespread credibility. "The essential complement to this broad ideological connection was a dense set of concrete [social] ties that enabled specific agencies and enterprises to construct joint projects at the sectoral level". "The 'embeddedness' created by these dense ties is central to the [...] developmental state" (Evans and Heller 2018: 2). Furthermore, the national purpose can bring together social and political forces in the interests of socially defined development-oriented agendas as a broad-based consensus may be necessary. Strategic partnerships between government, industry, and society can lead to a corporatist growth-oriented approach. Successful strategic planning should be based on a participatory model that includes various stakeholders since, in a democratic society, this proves vital to ensuring that the planning process is not hijacked by special interests. "Maintaining dense ties to entrepreneurial elites, while avoiding capture and being able to discipline these elites, is a defining feature of East Asian developmental states, distinguishing them from [the] less successful states [in the Caribbean]" (Evans and Heller 2018: 3).

Therefore, states should have a bulwark against special interest intervention designed to channel policy gains towards vested interests. Acting against harmful special interests requires technocratic insulation from external influence in both the planning and overview of the government's "national purpose" framework. A critically important difference between the successful developmental states of the post-war period and Caribbean governments is that the former typically had largely well-intentioned and "muscular" elites, whereas Caribbean nations have powerful anti-growth and anti-development coalitions (e.g., incompetent and corrupt politicians, union leaders, and business owners whose political and economic survival depends on keeping the *status quo*, and who fight hard to prevent the emergence of any potential challenge to their interests and power). If this barrier could be overcome, there is no other significant reason why it would not be possible to build strong national institutional structures in the Caribbean, although it would still require challenging the common desire of many politicians to exercise a high level of (often informal) control of institutions.

Clearly, the current economic difficulties, and fiscal pressures in particular, are making many governments more prepared to tackle difficult institutional issues than would have otherwise been the case. To be effective, Caribbean institutions would have to be embedded in productive networks with sections of industry, financial institutions, universities, and other key players. If these networks were successful, it would help to increase bureaucratic integrity, autonomy, and trust, a combination that would greatly improve policy quality and relevance. This combination of bureaucratic autonomy and social connectedness could provide the institutional basis for more effective, transparent, and accountable government involvement in Caribbean economies while remaining independent of vested interests and short-term political and societal pressures. In terms of economic policy objectives, however, the best developmental states were serious and effective in raising the quantity and quality of industrial investment. They focused almost exclusively on increasing productivity, growth, and exports, which served the strategic requirements of long-term economic transformation.

What differentiates the bureaucracies in the successful developmental states from the various planning institutions in other developing and less-developed countries is their relative independence, authority, technical competence and professionalism, and their ability to use both positive rewards and negative sanctions to influence decision-making in the private sector. Compared to these relatively strong and autonomous agencies, the weak and/or corrupt governments that are characteristic of many developing countries not only largely fail to deliver economically desirable interventions, but also inflict far too many politically motivated, inept, self-interested, or otherwise economically harmful policy actions. The relatively high level of public–private cooperation in the successful developmental states is not to be confused with the situation where, in some countries, the state has been largely captured by private interests, so that national goals effectively serve those private interests (this is true to varying degrees of Mexico, and many Latin American and Caribbean nations) (Leftwich 1995: 412).

So another important element of the successful developmental states is that they were not captured by corrupt special interest groups. But this relative autonomy requires the support of the political executive and reflects the fact that state power and autonomy were typically consolidated before other interests became influential, so that private economic interests were generally politically weaker, initially, than state power (Leftwich 1995: 416–17). Politicians have to provide the space for bureaucrats to make important decisions by ensuring that special interest groups do not deflect the state from its main development priorities and also legitimate and ratify the decisions taken by the bureaucrats. This combination serves when dealing with both civil society and local and foreign interests, giving the state the ability to shape advantageous economic relationships (Onis 1991: 115). It is important to note that the nature of the relative autonomy of the bureaucracy "[m]ust be seen as the product of a historical conjuncture of domestic and international factors. It is an autonomy embedded in the concrete set of social ties which bind state and society and provide institutional channels for the continuous negotiation and renegotiation of goals and policies" (Steinberg 1989: 74 quoted in Onis 1991: 123).

Jamaica was, until the early 1970s, the strongest economy in the Caribbean, with over two decades of high, sustained growth, averaging almost 5% per annum. The descent into political violence and corruption in the 1970s fundamentally damaged the economy, which showed little growth in the subsequent four decades. This pattern can also be seen in Trinidad and Tobago, considered the energy giant of the region, recording robust, sustained growth during the 1990s until 2010, with annual growth reaching over 14% in 2003. Excessive reliance on natural gas exports, failure to adequately develop resilience in other sectors, and weak or ethnically rationalized policies have resulted in negative GDP growth since 2016 (Macrotrends 2020). The latest CARICOM casualty is Guyana where political instability has emerged as the two ethnically polarized political parties fight to control the country's new found oil wealth (Kurmanaev 2020). By contrast, there was little or no ethnic conflict in any of the East Asian developmental states. Singapore, Hong Kong, and Taiwan avoided that trap by maintaining a strong state that could guarantee security and low rates of violence.

There are now three important questions. First, is there a characteristic East Asian development model? Second, is the developmental state model compatible with political liberalization and democratic forms of governance? Third, could—and should—this model be replicated in other developing countries, in particular, those (such as Jamaica, Trinidad, and Guyana) that have been on a low-growth path for over four decades? Would it help countries like Jamaica to recover their forward momentum? Although some governments have stated that their Asian model, broadly defined as combining economic liberalism with authoritative political action, is superior to that of the West, it is clear that, in some cases, the developmental state model was clearly a transitional phase. But in countries such as China, economic success has not (yet) been followed by political reform. The Party's legitimacy is based mostly on encouraging nationalism and delivering economic growth, so a failure on either count would expose it to increasing pressure. The question of whether China's model of political control can evolve into a Western-type democracy, therefore, emerges as a key intellectual inquiry for comparative political economy during the next decade.

It is important to note, however, that the Asian developmental state model was, in practice, the product of specific national and historical circumstances, which means that there may exist major constraints on the extent to which it could be replicated in countries which have quite different histories and cultures. What is important to learn from the East Asian experience of accelerated growth is how to take a more focused and strategic approach to development problems. Thus, the challenge for decision and policy makers in the Caribbean today is to devise forms of strategic development policy that are consistent with the norms of democratic accountability and with more limited concentration of state and private power than has been the case in the East Asian context (Karagiannis 2002; Karagiannis and Madjd-Sadjadi 2007).

All social and economic systems must adapt to change, or else perish. The East Asian states are evolving; the rise of their middle class, modern technological advances (e.g., green technological pathways), economic restructuring, and changes in the international environment have brought about changes in their elites, their coalitions of interest, and their institutions, as well as in their ideas of development. Hence, the developmental state model may evolve as sustainable development and green and blue growth paths could signal the beginning of a new phase for the developmental state model. It is clear, however, that the transfer of specific developmental state policies and strategies to new environments is unlikely to work in the absence of the political and institutional conditions that are required for their effective implementation.

What is an effective form of intervention for the Caribbean states today? Clearly, the state remains the most powerful institution to deliver social, developmental, and environmental goals. Although prediction has its risks, there are several characteristics of the modern developmental intervention, which emanate from the experiences in the post-war years in East Asia. First, governments have a key role to play in promoting social development, as the state has a primary responsibility to provide adequate social services and promote the well-being of citizens. This emphasizes the value of social programmes that attempt to break the cycle of poverty and thereby reduce inequality, which may ultimately prove to be a more durable basis for long-term prosperity (Karagiannis and Madjd-Sadjadi 2007). Therefore, well-planned social programmes should be an important component in any development strategy because they meet the basic needs of a wide spectrum of the population (MacEwan 1999: 176). Social programmes also have implications in terms of other goals of national development, as good programmes on education and health can increase *per capita* productivity and thereby support economic growth (MacEwan 1999: 177). This depends heavily, however, on the way that social programmes are constructed, as the expansion of social programmes per se does not necessarily lead to economic growth or other desirable ends. Even the best shaped social programmes (in terms of their impact on productivity) are likely to have limited impact on economic growth or competitiveness unless the overall development strategy also contains a growth-oriented macroeconomic framework, training programmes, and measures to encourage the private sector to create higher-quality jobs (MacEwan 1999: 195).

In the Caribbean, there is a particular need to improve the functional relationship between the government and the private sector. A number of Caribbean governments have been effectively captured by powerful vested interests (Trinidad & Tobago and Guyana, for example, where violence and weak institutions are also due to competing ethnic groups—political and elites), and are therefore unable to decisively promote necessary structural changes and economic reforms. The politicization of most decisions, in conjunction with the weakness of the civil service, means that there are major constraints on local developmental policies. The developmental state strategies and policies by Caribbean governments would therefore require strong institutional structures with a substantial measure of operational autonomy to manage the critical interactions between state and industry. This would allow a genuine complementarity between public and private goals, enhance the operation of markets, and create opportunities that would not otherwise exist.

A critically important difference between the successful developmental states of the post-war period and the Caribbean nations today is that the former typically had powerful but largely well-intentioned elites, while the Caribbean has powerful anti-growth and anti-development coalitions whose political and economic survival depends on keeping things as they are. There are, for example, incompetent and corrupt politicians, union leaders, and business owners who will fight hard to prevent the emergence of any potential challenge to their monopoly on power. If this barrier could be overcome, there is no other significant reason why it would not be possible to build strong institutional structures in the Caribbean, although it would still require challenging the common desire of many politicians to exercise a high level of (often informal) control of institutions. To be effective, these institutions would have to be embedded in productive networks with sections of industry, financial institutions, universities, and other key players. If these networks were successful, it would help to increase bureaucratic integrity and autonomy, a combination that would improve policy quality and relevance.

This combination of bureaucratic autonomy and social connectedness (what Evans [1995] calls "embedded autonomy") could provide the institutional basis for more effective, transparent, and accountable government involvement in Caribbean economies, while being independent of vested interests and short-term political and societal pressures. This approach allows "considerable autonomy in determining the mode of operation, and adjusting it as experience accumulates". The main objective is "a dynamic economy rather than sticking to a set of rigid rules imposed by a central bureaucracy". We must avoid squandering people and resources over a whole range of bureaucratic activities (Cowling 1990: 25). In the Caribbean, there have been attempts to create approximations of such productive networks through, for example, the creation of committees to monitor issues of transparency and oversight in key productive, government-controlled sectors. But these attempts have met with varying degrees of success in terms of operational autonomy.

Traditional government bureaucracies were largely focused on day-to-day management and control, required little conception of strategy, and rarely undertook any serious forward planning. Indeed, strategy of any kind would have been considered "political" (Hughes 1998: 149). However, stronger institutions with a clear strategic planning role are a necessary part of any attempt to introduce greater competence and a more long-term perspective to development policy in the Caribbean, a combination that would be more favourable to productive investment and sustained economic growth. This would also allow the state to take a genuinely proactive role, harnessing market forces to deliver the goals set by government at various levels, local and national (Cowling 1990: 32). The role of strategic planning is to focus on desired goals (rather than operational details) in order to give direction and purpose to public organizations; without this underlying strategy, policymaking is often without coherent direction. The planning process is often more important than the plan itself, as the use of long-term parameters allows public institutions to develop a shared vision for the future (Cowling 1990: 16–17).

A system of accountability is still required by Caribbean governments, as the two forms of accountability, political and managerial, are tightly related. As Hughes (1998: 225) states, "the system of accountability is what ties the administrative part of government with the political part and ultimately to the public itself". In the traditional bureaucratic administration there is some accountability, which relies upon the formal hierarchical structure. But this is typically accountability for avoiding errors, rather than achieving any actual outcomes (Hughes 1998: 233). Improving the idea of accountability by relating it to the achievement of actual goals would therefore be another obvious goal of any move towards more effective developmental intervention. This would involve making relevant institutions responsible for the effective execution of strategic plans and monitoring the progress of these plans. As Cowling puts it:

> . . . [I]nstitutions can formalise the commitment to such [strategies], and their structure, procedures and personnel can act to ensure that such commitments cannot easily be reversed, but they are simply ratifying plans already established. The history of planning [in Caribbean nations] shows how fragile was the commitment, despite the creation of many new institutions, [and the lack of teeth of these institutions was obvious]. With clear goals, and a determination to pursue them, institutions with teeth should be forthcoming [in the Caribbean]. (Cowling 1990: 23)

All too often, public sector reforms and capacity-building programmes have been introduced in many Caribbean countries without the benefit of systematic and disciplined diagnoses of their institutional capacities. This has resulted in wasted spending, inadequate levels of competence and skill, ineffectiveness, and performance shortcomings. In contrast, the economic success of Japan and NICs could not have been achieved without the decisive role of their competent "technostructures" in economic and social planning. Indeed, an important feature of these technocracies has been their technical competence; many of the top officials of Japan and East Asian countries received advanced training abroad. It would, in principle, be possible to build a more competent technocracy, and the same underlying knowledge-based strategy could then switch to place the emphasis on increasing the skill base in other areas, including innovation hubs, the capacity to develop green and blue infrastructure, make more effective use of technology including technology generation, acquisition and transfer, higher education and research, continuous professional development, and other factors that serve as growth accelerators.

In all such cases, it is important to connect the necessary intellectual, managerial, and fiscal resources to maximize the synergies. Strategic human resource management and planning coupled with investments in human resource development (i.e., high quality and timely education, training, and the continuous development of scientific manpower) should therefore be linked to a clear effort by Caribbean governments and institutions to improve the quality and effectiveness of state actions, including a commitment to pay competitive wages for highly competent, trustworthy, and morally committed technocrats. It is essential, however, that Caribbean governments should eliminate political patronage since, in a culture of patronage, key posts are often filled with party loyalists or political allies, and this can very effectively undermine national development.

Developmental intervention also requires better use of resources. This involves directing resources to emphasize those programmes that most assist the attainment of strategic objectives. It also involves more state spending on key infrastructure and the other modern factors of growth and competitiveness, and less spending on non-essential kinds of government expenditure (like "social programmes" that channel patronage). It is now possible to move most government services online, which creates scope for a large reduction in public sector staffing levels. These reductions could still be based on strategic human resource management and planning so that a relatively smaller but much more expert civil service replaces the unwieldy and inefficient bureaucracy. This would help to raise competence levels and also release resources for higher levels of green investment.

Public investments in knowledge-building, technological innovation, training, and research can be oriented to the needs of the business sector with a view to crowding in private investment. This could be at least partly financed by eliminating the large array of tax waivers and subsidies now used to support politically favoured businesses. Caribbean governments, however, still have to find ways to ensure that the best business practices are adopted and require firms to invest in green production facilities, skills training and upgrading, and critical science and technology initiatives. The nations which do best in the global arena are those which manage change competently and encourage their private sector to invest in new opportunities. The capacity to manage change is the best way to ensure stability and prosperity.

The effectiveness of this model would also depend on the coherence of state policies, which is clearly going to be difficult to maintain when important parts of the state are wedded to pork barrel politics and beholden to specific interests. Effectiveness in the Caribbean, therefore, will be highly dependent on the degree of relative institutional autonomy from powerful world organizations and transnational corporations as well as the surrounding social and political structures and pressure (Wade 1990: 375). In the course of a fundamental redirection of this kind, simply matching developmental state policies to existing political institutions is very unlikely to work and probably would be counter-productive. So any such change would require a deep commitment to fundamental improvements in policymaking, while the institutional structures would have to generate value by providing continuity, consistency, and commitment to the direction and pace of green development (Ahrens 1997: 119, 126). Without such commitments, effectiveness, competence, capacity, and professionalism, developmental state policymaking will probably flounder on short-term expedients; the power of existing interests; the conservatism, ineffectiveness, and inefficiency of the civil service; or the resistance of 'ceremonial' organizations and institutions (Cowling 1990: 23). The rest of this chapter offers developmental state-led green and blue growth strategies for Caribbean economies.

4 Developmental State-Led Green and Blue Growth Strategies for Caribbean Economies

Given that Caribbean tourism, the region's largest industry, has experienced significant growth and has generally performed well during the last 30 years, this last section seeks to offer strategic policies aimed at boosting the industrial competency and market development potential of local firms. Yet, national production growth may demand politico-institutional changes, which may also require regime changes in order to bring into power new forward-looking entrepreneurs and manufacturers, economic technocrats, and progressive intellectuals. The development challenge for a Caribbean nation's policymakers is to devise forms of industrial policy that are consistent with the norms of democratic accountability.

An important but highly controversial issue today is to bring "the nation state back to business", and simply concerns the economic impact of development policy vis-à-vis other industrial growth determinants. This calls for more engaged states that are also more effective, accountable, and results-oriented. Our line of argument is based on a regional-national-local interplay, with thorough, effective, and technically proficient action by the national government being the most important piece of the development puzzle. It bears remembering that the developmental state springs from a view which accepts that government plays an important role in economic restructuring, increasing the quality and quantity of industrial investment, and fostering links with civil society. Government policy and influence would have to be decisive factors, and strategic planning should be limited to selected policy arenas. In order to forge lasting relationships between government and industry, government strategic planning must embrace the creative dynamics of the market as much as the market must adapt to the strategic plan—in this way, effective coordination may be realized.

In contrast to state interventions which are consonant with the *market failure* view, a government attempting to actively boost industrial growth needs to have a dominant developmental orientation. Consequently, the basic vision of a Caribbean-type twenty-first-century developmental state remains compelling as a coherent and capable state apparatus can boost green and blue growth activities to produce forward-looking investments that can enhance productivity, grow incomes, and lead to increased well-being in the region. For this reason, and perhaps for others, the construction of a green production-based approach to economic development and a much sharper focus on strategic industrial policy should be important parts of a "Caribbean-type" developmental state, are seen to be necessary to resolve structural problems, and offer concrete alternative solutions to greening Caribbean economies. In effect, strategic industrial policy targets and centres around key strategic green and blue sectors, which can be expected to fuel local diversification and self-sustained growth. By recognizing differentiation of new and promising activities, particular branches and industrial sectors, thorough technically proficient policy can take care of the necessary human, material, and financial requisites and thus become effective.

Tourism—although a low skill-low technology sector—is one of the most important economic activities in the Caribbean, contributing a third to half of GDP in most countries. The fundamental environmental resources used in Caribbean tourism are basically the "sun, sand and sea". Whether tourism is based on stopover visitors or cruise ships, Caribbean countries rely on their natural environment as the main lure to visitors, as well as an important source of welfare for their own citizens. However, the resources upon which the tourism industry is based are very fragile. Except for the sun, each of these resources—as well as supporting resources such as potable water—are subject to damage and depletion. Threats to environmental resources used by the tourism sector can arise both within and outside the sector.

Given the relative weakness of most Caribbean governments today, it would be sensible to limit strategic intervention to those parts of the Caribbean economies where government intervention could have the greatest positive impact on dynamism and sustained economic growth.

Subsequently, it is important to divide consideration of key issues related to the structure of Caribbean economies into three sections: (1) issues influenced by policy intervention and general policy issues; (2) issues influenced by specific industries or sectors; and (3) market-driven issues. One important long-term goal would be to bring about an improvement in the competency and efficiency of local firms, in the level of technological infrastructure they rely on, and in the quality of workmanship and service, so that more and more activities may become increasingly competitive. It would not be sufficient to establish small, technologically modern export sectors, for example, without a plan for using them in a programme to upgrade the overall skill level in the workforce.

In some sectors, the region already has a strong basis on which to build (tourism and hospitality, entertainment, agro-processing, and food production). These sectors, as well as modern activities of high potential—like those based on alternative or renewable forms of energy and ocean ecosystem services—require significant investment, rejuvenation, and repositioning and have to address a number of serious economic, social, and environmental issues simultaneously. Provided that the immediate problems are solved, the targeted green and blue sectors are clearly capable of considerable further expansion. In fact, the mutually beneficial relationship between different types of tourism and planned industrial activities of high growth potential and achievability can provide the foundation on which alternative national development strategies can build. Once the priorities are right, and with effective "policy spillovers", natural resources will increasingly be allocated efficiently, production, productivity, and profitability will increase, and the propulsive and dynamic green and blue sectors will become increasingly attractive to the private sector.

In addition, the approach used here seeks to link the process of growth and change in various sectors with that in the economy as a whole. The growth process in conjunction with higher levels of tourist arrivals are expected to lead to a widening of local markets, which in turn can bring about better transportation and communications systems. After resources have been developed and/or put to use, better industrial capabilities will broaden the Caribbean production base, will anchor the growth of domestic knowledge and technological capacity, will enhance green financing options, and will provide sufficient stimulus to the mobilization of resources of all kinds and the inducement to invest. But domestic industrial capabilities formation—which encompasses knowledge, financial, and technical supporting institutions—is a sequential and cumulative process that can flourish in an environment of political and macroeconomic stability.

There is no need for vast bureaucratic machinery and procedure because the approach is clearly entrepreneurial. Such a radical approach needs to: emphasize substantive national development goals while being mindful of the constraints that the international political economy imposes; consider key relevant aspects of the East Asian trajectory towards boosting green production expansion while making the necessary modifications to reflect Caribbean history, culture, social psychology, and politico-institutional arrangements; acknowledge the importance of authoritative and focused government agencies in overseeing the course of Caribbean industrial resurrection; and determine the quality and impact of policy intervention. A successful implementation, however, will require wide consultation, broad consensus, trust, government intervention of high quality, determination, realism, commitment, and special emphasis on green production-oriented growth.

The proposed framework takes into account dynamic relations among a number of "stylized facts": local resources, capital, social structure, technology and skills, scale and scope, institutions, history, culture, and social psychology. Such a pragmatic policy approach can successfully contribute to long term, supply-side initiatives aimed at creating or promoting selected sectors and prioritized activities, and can create economies of scale and scope, conditions and opportunities conducive to faster growth of existing and incoming enterprises (a "big push"). Economies of scale and learning will bring about multiple effects on, and changes in, the structure of local economies.

The object, of course, would be to increase value-added to the emerging sectors and strengthen forward and backward linkages, which would then be capable of spilling their expansionary forces into other sectors and activities: the support and development of indigenous resources, firms and industries; the maximum utilization of investment (mainly in relevant R&D, innovation and skills to support new distinctively Caribbean green and blue industries); the removal of constraints on both the demand and supply sides, which are imposed by the narrow size of the local markets and poor manufacturing base of Caribbean nations; an improvement in the range of services available to people and to industry (e.g., biodiversity, maritime transport, coastal protection, waste disposal, and carbon sequestration); the exploitation of economics of scale and scope; the application of productivity-enhancing production methods and techniques so as to raise industrial efficiency and competitiveness; the rejuvenation, restructuring, diversification, and strategic global repositioning of Caribbean economic activity; and, the capacity to form the foundation of a green or blue economy that safeguards sustainability and social and economic equity while correcting the region's tendency towards external disequilibrium and high dependency on foreign economic activity.

Such a green alternative approach requires considerable discourse to ensure that policy goals are clearly identified. Greater local ownership in the Caribbean could help improve green development outcomes by boosting indigenous industrial competency. In some sectors (i.e., different types of tourism, entertainment, local food and beverage, green technologies and renewable energy, as well as other production lines that can utilize these alternative energy sources), Caribbean economies have a good foundation on which indigenous productive initiatives can build. Yet, the creation of winners through industrial policy depends on the development situation of the nation. Particular industrial targeting goals may differ from one Caribbean country to another, as industrial priorities may need to be different. In the Bahamas, for example, industrial policy goals should first emphasize an internally propelled agro-industrial development given the country's negligible local production. This local production growth would be able to diversify the economy, create strong linkages with other sectors and activities, and improve the country's unsatisfactory trade performance as an extension. In Jamaica, industrial policy goals should centre on augmenting agro-industrial production lines and diversification, along with stronger linkages with other sectors and activities, and on producing competitive intellectual property, which is today a major economic driver globally. In Trinidad, industrial policy goals need to take the country's tourism performance into account and emphasize planned tourism growth, which would not only boost local sectors and activities but also enable national authorities and policymakers to control economic, environmental, cultural, and social outcomes.

While consumption is the vehicle through which members of society achieve present satisfaction and happiness, it is investment that crucially determines the future wealth of a nation. Thus, greater emphasis should be placed on strategic planning and green investment (e.g., green and blue bonds)[2] that will lead to future prosperity. Planned spending on green technological innovation and skill formation will provide the industrial requisites to thoroughly support emerging green and blue sectors and activities. In so doing, government policies—whether related to human and physical capital investment; fiscal and financial policies; research and development (R&D) incentivization and technology transfer; or trade measures—all need to be aligned with these strategic goals. However, the effectiveness of such comprehensive strategies and policies would be evident in the long term. Such knowledge-based strategies can be spawned by the University of the West Indies, and would emphasize infrastructural development, machinery and capital equipment, technical change, green technology generation and transfer, continuous development of scientific manpower, a higher level of education and cutting-edge research, training/retraining courses, and learning-by-doing.

At the regional level, given the Caribbean countries' traditional disadvantages and diseconomies of scale, scope, and space as well as their susceptibility to volatility of income and vulnerability to natural and environmental disasters, "growth pole" strategies are a pragmatic response to constraints associated with small island economies. By supporting the broad base of green local technological accumulation and innovation, growth pole strategies would create greater external economies and economies of scale, thereby dealing with issues of scale and scope. The existence of these external economies would create conditions and opportunities conducive to faster growth of existing and incoming industry in the Caribbean and would make the growth centres more attractive for new industrial development. Similarly, Caribbean states may emphasize "clusters" as important engines of green industrial growth. More importantly, growth pole strategies in the Caribbean would be capable of spilling their multiple benefits into depressed neighbouring areas within the region.

Moreover, green growth has already begun to respond to, and internalize a number of signals—economic, social, technological, ecological,

[2] Also, Block (2014: 3) advocates for "an expanding network of non-profit financial institutions that would redirect [...] savings to finance clean energy, growth of small and medium-sized enterprises, and [green] infrastructure". In the Caribbean, the mission of such financial intermediaries would be to facilitate and finance green and blue economic activities in an island or a particular geographical area.

cultural, and institutional—which stem from the modern global environment. With cooperative regional arrangements but national approaches to green and blue growth, Caribbean islands can share the expenses of building marketing intelligence systems, information technology networks, promotion, and public relations campaigns. These cooperative arrangements and united efforts can considerably increase the strength of the bargaining power of the entire region in important green growth policy areas. However, current world happenings pose tremendous challenges for the Caribbean and beyond, thus, calling for further research and scholarly contributions on the geopolitical landscape of global mobility across many countries around the world.

5 Conclusion

Development in any Caribbean nation is a dynamic process that cannot occur in a vacuum. Hence, policy choices must be made with a clear understanding of the role of the political, social, cultural, philosophical, religious, historical, and economic realities and linkages in the broader society as well as an appreciation of the likely reaction of international institutions and trading partners. Central aspects of the proposed green developmental state approach in the Caribbean context are: a strong ambition to develop; priority being given to expanding green production; state-provided guidance and support to emerging targeted sectors; an economic policy bureaucracy within a political network that offers sufficient space for initiative-taking and effective operation; and a "strong" "pilot agency" with professional, competent, and morally committed members as the brains for carrying out green policy interventions.

As it is impossible to specify the results of such a process *ex ante*, the crucial task is to discover where intervention is needed and what type of action can bring forth the greatest response. This is the reason why the developmental state should promote inclusive and equitable green growth while taking care of the necessary political, institutional, and financial support. Although the international political economy may constrain developmental paths, policy space and choices, the accomplishment of these green development goals clearly requires better state action, "functional finance" and green investments, which are most likely to be achieved via developmental state policies. Therefore, a rational and workable green developmental state line of argument is advanced here as a necessary mechanism for support of selected green and blue sectors and activities in Caribbean economies while leaving space for stimulating further social and political advancement.

References

Ahrens, J. (1997). "Prospects of Institutional and Policy Reform in India: Toward a Model of the Developmental State," *Asian Development Review*, **15** (1), 111–46.

Amsden, A. (1989). *Asia's Next Giant*, New York: Oxford University Press.

Block, F. (2014). "Democratizing Finance," *Politics & Society* **42** (1), 3–28.

Evans, P. (1995). *Embedded Autonomy: States and Industrial Transformation*, Princeton, NJ: Princeton University Press.

Cowling, K. (1990). "The Strategy Approach to Economic and Industrial Policy," in K. Cowling and R. Sugden (eds), *A New Economic Policy for Britain: Essays on the Development of Industry*, Manchester: Manchester University Press, pp. 6–34.

Evans, P. and P. Heller (2018). "The State and Development," WIDER Working Paper 2018/112, United Nations University-WIDER, September.

Focus Economics (2020). *Trinidad and Tobago Economic Outlook*. https://www.focus-economics.com/countries/trinidad-tobago

Hughes, O. E. (1998). *Public Management and Administration*, 2nd ed., London: Macmillan.

Johnson, C. (1982). *MITI and the Japanese Miracle: The Growth of Industrial Policy, 1925–1975*, Stanford: Stanford University Press.

Karagiannis, N. (2002). *Developmental Policy and the State: The European Union, East Asia, and the Caribbean*, Lanham, MD: Lexington Books.

Karagiannis, N. and Z. Madjd-Sadjadi (2007). *Modern State Intervention in the Era of Globalisation*, Cheltenham, UK and Northampton, MA: Edward Elgar Publishing.

Kurmanaev, A. (2020). "Oil Bonanza Plunges Guyana Into Political Crisis," New York Times, March 5. https://www.nytimes.com/2020/03/05/world/americas/guyana-elections-oil.html#:~:text=Oil%20Bonanza%20Plunges%20Guyana%20Into%20Political%20Crisis%20The,has%20deepened%20ethnic%20divisions%20and%20exacerbated%20tensions%20

Leftwich, A. (1995). "Bringing Politics Back: Towards a Model of the Development State," *Journal of Development Studies*, **31** (3), 400–27.

MacEwan, A. (1999). *Neoliberalism or Democracy?* London: Zed Books.

MacroTrends (2020). Trinidad and Tobago GDP Growth Rate 1961–2020. https://www.macrotrends.net/countries/TTO/trinidad-and-tobago/gdp-growth-rate

Nester, W. R. (1991). *Japanese Industrial Targeting*, London: Macmillan.

Onis, Z. (1991). "The Logic of the Developmental State," *Comparative Politics*, **24** (1), 109–26.

Singapore's Blue Economy. http://eascongress2018.pemsea.org/wp-content/uploads/2018/12/S7.3-2.-Singapores-blue-economy-innovations_SWYin.pdf

UNDP (2010). Responding to Climate Change in Small Island Developing States. https://sustainabledevelopment.un.org/content/documents/960SIDS_Flyer_SEPT_27_09[1].pdf

UN Sustainable Development Goals Knowledge Platform. Small Island Developing States. https://sustainabledevelopment.un.org/topics/sids

Wade, R. (1990). *Governing the Market: Economic Theory and the Role of Government in East Asian Industrialization*, Princeton, NJ: Princeton University Press.

China's Ecological Civilization: From Contradiction to Synthesis

Vincent Huang

1 Introduction

The primary purpose and contribution of this paper is to examine China's environmental reality and the evolution of China's environmental policy after the country's 1978 economic reform. I argue that in attempting to build an ecological civilization, China's environmental policy has evolved from contradiction to synthesis. Echoing my other essay on China's alternative political economy (which is defined by the dominant, evolving, and developmental role of the Chinese state in economic and social processes), I argue that the Chinese state has been and will be shaping China's environmental landscape more responsively and effectively now and into the future.

I divide the paper into four parts. Part I briefly surveys the reality of China's dual economic and environmental crises that have led to contradictions in China's environmental policies. Part II examines the debate

V. Huang (✉)
University of Denver, Denver, CO, USA
e-mail: Yijiang.Huang@du.edu

N. Karagiannis and J. E. King (eds.), *Visions and Strategies for a Sustainable Economy*, Global Institute for Sustainable Prosperity,
https://doi.org/10.1007/978-3-031-06493-7_13

between mainstream environmental economics and heterodox ecological economics regarding the relationship between the economy and the environment and the extent to which it is applicable to China. Part III examines China's era of contradictory environmental policy and ineffective environmental governance from 1978 to 2005. Part IV elaborates the evolution of China's environmental policy from contradiction to synthesis since 2005. While it is difficult to present the totality of China's environmental policy climate, I hope to demonstrate to readers that China's environmental policy today is in a historically stronger political and institutional position to take advantage of creative public policies such as a Green Job Guarantee to engineer environmental sustainability while achieving and maintaining full employment.

2 China's Dual Crises of Unemployment and Environmental Degradation

China's mutually reinforcing and exacerbating economic and environmental challenges have led to contradictions in China's environmental policy. At first glance, China does not seem to have a severe unemployment problem. Its official unemployment rate has been around 4% for decades, irrespective of business cycles. As far back as in the mid-1990s (when millions of state-owned enterprise workers lost their jobs due to the economic restructuring) and as recent as in the 2008 global financial crisis (when China's exports industry took a major hit), China's official unemployment rates remained at 3.1 and 4%, respectively. The answer to such statistical absurdity seems to be the political concerns that had impacted the impartiality of the National Bureau of Statistics of China.

Upon examination, one realizes that China's unemployment rate is not internationally comparable by design. The National Bureau of Statistics of China computes the country's official unemployment rate by using the registered urban unemployment rate, which is different from the international standard LFS-based (Labor Force Survey) unemployment rate. China's urban registered unemployment rate has two features that tend to drastically understate the actual state of unemployment.

First, the term, "registered," means that one cannot be officially unemployed in China until s/he reports to the government. In comparison, the LFS-based unemployment rate, due to its very nature of household survey, includes those unemployed that have not reported to the state bureaucracy. Second, the term, "urban," means that only urban Chinese[1] are included in the computation of the unemployment rate. Rural Chinese, which account for 57.65% of China's total population as of 2018,[2] are excluded from China's computation of its official unemployment rate.[3]

Hence, China's surprisingly low and stable unemployment rate is a product by design. In recent years, China has begun to reform its unemployment statistical reporting. Since 2018, China has published the LFS-based unemployment rate for urban population, and the result is about one percentage point higher than the corresponding urban registered unemployment rate. If fully adopting the international standard,[4] however, China's true unemployment rate exceeds 8%, an estimate provided by the Chinese Academy of Social Sciences.[5]

The estimate provided by the Chinese Academy of Social Sciences is probably on the conservative end. But even assuming a 5.2% urban unemployment rate, together with a 466.74 million urban labor force, paints a very different picture for the Chinese economy—a mega economy that grows over 6% in 2019 but still has an unemployed population of 24.27 million (National Bureau of Statistics of China 2019). Whether it is the sheer size of the unemployment or the 280 million peasant workers that are precariously employed in the cities, China continues to face a serious unemployment challenge that threatens its economic, social, and political stability ever since its second phase of economic reform in 1993.

The economic and social pressure to maintain employment and promote economic growth has for decades contributed to the Chinese

[1] Each Chinese citizen is classified as either rural or urban by birth in the Household Registration Book, a legacy from the Socialist China. For details, see Chapter 2.

[2] As classified by China's Household Registration System.

[3] It is assumed that rural Chinese can always rely on peasantry in rural China for a livelihood, and thus can never be actually out of work.

[4] Computing LFS unemployment rate for the entire working age population, regardless of rural or urban status.

[5] The Chinese Academy of Social Sciences is a premier academic research organization affiliated with the State Council of People's Republic of China.

state's reluctance to seriously tackle environmental issues. Various environmental indicators illustrate the magnitude of China's accumulated environmental crisis. The most recent "Environmental Performance Index" (EPI) ranks China 120th out of 180 countries in 2018, which puts China behind Russia (52nd) and Brazil (69th) and only ahead of India (177th) among the BRIC nations (Yale Center for Environmental Law & Policy 2019). Most of the countries low in the ranking are nations that face more direct political or economic emergencies. In the case of China, however, its relatively low position in the EPI ranking reflects the contradictions in the country's environmental policies for the most years since the 1978 economic reform, such as the late introduction of policies to reduce fossil fuels dependency in the early 2000 (Piovani 2017). Hence, as one of the fastest growing economies, China has experienced rapid environmental degradation in the past three decades, with an environmental sustainability index now near the bottom of the world.[6] China's environmental crises are particularly evident in the areas of air pollution, water contamination, and land erosion.

The seriousness of China's air pollution is well-documented. As of 2016, China has the fifth highest death rate in the world in terms of the percentage of premature death as caused by air pollution (McCarthy 2018). According to Greenpeace, "in 2017, average PM2.5 [air pollutants that seriously affects human health] concentrations in Beijing exceeded World Health Organization (WHO) air quality guidelines by nearly sixfold, and by almost fourfold in Shanghai." Furthermore, "in the 338 cities for which monitoring data is available, average PM2.5 concentrations amounted to 4.3 times the WHO guideline" (Greenpeace 2018). Ozone depletion is also a major concern in China, as "average ozone exposure in China rose 17% between 2014 and 2017, an increase which caused an estimated 12,000 premature deaths per year" (Greenpeace 2018). China's air pollution requires immediate public response.

Meanwhile, China's water contamination is also pressing. Population growth and industrialization, coupled with loose environmental regulation, have exacerbated China's water contamination and shortage. As

[6] According to the Environmental Sustainability Index composed by the Socioeconomic Data and Applications Center at Columbia University.

shown in a 2014 Chinese government report, 59.6% of China's underground water is classified as "bad" or "very bad." By 2017, China had established 12,226 monitor sites to monitor the quality of surface water across the country. The Institute of Public and Environmental Affairs reports that, of these 12,226 sites, "just 35% have water of good quality, another 32% are suitable for water supply, 20% are suitable for industrial or agricultural use – but not for human contact – and 13% are useless. Even in Shanghai, one of China's richest, most environmentally-aware and modern cities, 52 out of 65 monitoring sites have water not suitable for human contact" (Webber 2017). The result is that "two-thirds of China's 656 cities (with 390 million people) suffer shortages of water for domestic and industrial use" (Li 2016).

China is additionally plagued by land erosion, especially due to deforestation and desertification. It is estimated that illegal logging consumes around 5000 square kilometers of virgin forest every year in China. According to China's Ministry of Ecology and Environment, the use of disposable chopsticks alone consumes up to 1.3 million cubic meters of timber every year. Furthermore, deforestation compounds desertification, as "the rate of desertification in China is around 900 square miles a year, an area nearly the size of Rhode Island, with an area the size of New Jersey becoming desert every five years." However, it must be noted that, compared to air and water pollution, China's deforestation and desertification have been much better tackled in recent years. The public policies underlying China's massive reforestation and anti-desertification efforts are precursors to the Green JG program that I'd like to propose.

China's dual economic and environmental crises, as highlighted above, are mutually reinforcing and exacerbating. China's 8% unemployment rate mandates an ever-growing economy to absorb the excess labor supply and maintain economic and social stability. However, doing so may further deteriorate the environment. On the other hand, China's worsening environmental crises call for a structural economic reform away from heavy and polluting industries. But then again, doing so may exacerbate unemployment and erode social stability. It is this economic-environmental tradeoff that has led to contradictions in China's environmental policies following the country's 1978 economic reform, which has rendered China's pursuit of a green civilization a distant future for the most years after the reform.

3 The Economy and the Environment in China

After presenting China's reality of dual economic and environmental crises, I turn to economic literatures to survey the debate between mainstream environmental economics and heterodox ecological economics regarding the relationship between the economy and the environment and the extent to which it is applicable to China. I argue that mainstream environmental economics has explained the contradictions in China's earlier environmental policies albeit with a flawed analysis. Meanwhile, heterodox ecological economics seems to be consistent with China's instrumental environmental policies in more recent years.

Mainstream environmental economics is appropriately a subfield of neoclassical economics in that it retains most of the latter's theoretical foundation. Under normal circumstances, the intersection of an upward-sloping supply curve and a downward-sloping demand curve determines a set of equilibrium quantity and price in a goods market that most efficiently allocates scarce resources according to tastes and preferences of market participants. The existence of environmental degradation in the face of market allocation therefore entails two broad categories of explanation within mainstream environmental economics.

The first explanation is that environmental degradation occurs because a clean environment does not reflect the tastes and preferences of existing market participants. This argument can be best illustrated by the environmental Kuznets curve (thereafter EKC), which hypothesizes that as economic growth occurs, environmental degradation worsens at first and then, after per capita income reaches a certain level, improves as economic growth continues to develop. Thus, this model has been frequently applied to justify environmental exploitation in China and other developing countries where per capita income is assumed to be lower than the turning point level of per capita income in the EKC (Stern 2004).

The economic interpretation of the EKC is that at the early stage of economic development, people have lower per capita income and are, therefore, more willing to exploit the environment for economic benefits. As people become richer, however, their taste and preference structure changes to prefer a cleaner environment.

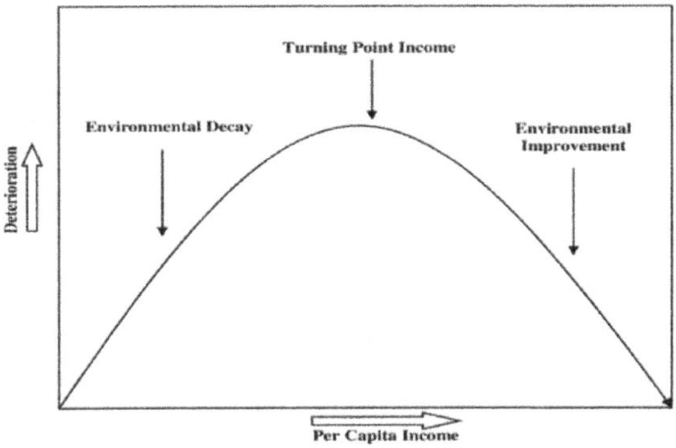

The implication of the environmental Kuznets curve seems ironic: the best way to resolve environmental issues is to promote quantitative economic growth, which is what caused environmental degradation in the first place. Numerous critiques have been made of the EKC argument, ranging from its problematic assumption to sustainable economic development to its weak econometrics and problems in pollutant specification (Yandle et al. 2004). Nonetheless, the EKC is consistent with mainstream environmental economics, which suggests that environmental degradation in the face of efficient market allocation is a matter of revealed tastes and preferences.

The other explanation is the market failure argument. In this case, it is possible that market participants do value the environment. However, the problem is that the market has failed to reflect people's environmental tastes and preferences in the case of a Pigouvian negative externality (Pigou 1920). Such a negative externality is the failure of market participants to internalize all relevant external costs in their decision-making, and thus may adversely affect other people who bear the cost (Pigou 1920). A negative externality thus leads to overproduction.

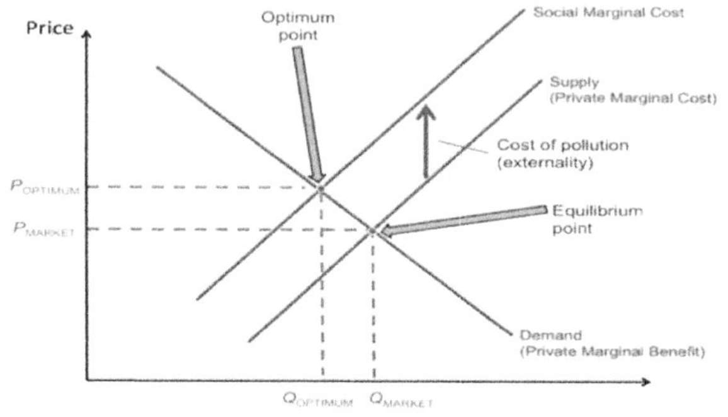

Since environmental degradation is due to market failures, the solution is to correct them via proper state intervention. This is done by estimating the excluded external costs and augmenting them to the firm's private costs. This will ensure that the firm's level of production would reflect the optimal utilization of resources according to the society's tastes and preferences. Public policies should, therefore, focus on the imposition of "Pigouvian taxes." After the elimination of negative externalities, a society achieves its optimal level of pollution.

I argue that mainstream environmental economics has explained the contradictions in China's earlier environmental policies albeit with a flawed analysis. To give the EKC thesis credit, China has indeed sacrificed its environment for economic gains for the most part of its post-1978 economic reforms. These sacrifices ranged from the reliance of the economically cheap but environmentally costly coal as the primary source of energy generation to the loose implementation of environmental protection laws in order to establish China's manufacturing dominance globally. However, to believe in the self-adjusting implication of the EKC argument would be misleading and dangerous. It is difficult to argue that the numerous environmental crises in China today, such as the "Beijing Smog" and the desertification of arable land, reflect the tastes and preferences of the Chinese people or the Chinese state. China's environment had continued to worsen even as environmental concerns replaced land dispute as China's major cause of social unrest.

The correlation between economic growth and environmental degradation as hypothesized in the EKC is not necessarily causative. Perhaps the United States reversed some of its environmental degradation not by achieving a per capita income level so magical that Americans no longer exploit the environment for additional incomes. Instead, the United States might have reversed some of its environmental degradation by exporting pollutant industries to other countries (such as China) that could produce goods more cheaply and thus more profitably. This suggests that we must not expect environmental degradation in China to fix itself with mere more economic growth. In the case of China, as the Chinese state is determined to maintain its manufacturing dominance, it cannot eliminate environmental degradation by simply exporting its manufacturing sector abroad and solely focusing on developing the service sector. While countries like Vietnam, India, and Bangladesh continued to absorb the low-skill and low-cost manufacturing jobs from China, China must actively address any existing issues with its environment.

Regarding the negative externality argument, China has indeed been following its policy prescription by enacting various kinds of pollution taxes ever since 1982. However, it is inconceivable that these pollution taxes had led to an optimal level of pollution in China. Perhaps China's pollution taxes could be increased to a much higher level to be effective. But even then, the concern is that it could discourage production and lead to more unemployment, causing economic and social instability. Addressing environmental degradation while maintaining a stable and growing economy thus becomes a paramount task for the Chinese state to engineer a green civilization while actualizing the "Chinese dream."

However, even if we put external consistency aside, both the environmental Kuznets curve argument and the Pigouvian externality argument are problematic in their fundamental assumptions and approach. Both share the implicit logic that the environment is like any other goods in the economy—we should consume/pollute the environment so long as the marginal benefit of polluting outweighs the marginal cost. Both EKC and Pigouvian externality are different forms of cost–benefit analyses that assume the superiority of monetary efficiency over environmental and social efficiency. Mainstream environmental economics is therefore the economics of the environment that subjects the environment to the economy and environmental problems to cost–benefit analyses.

However, its view of the relation between the economy and the environment is an example of a "fallacy of misplaced concreteness" (Daly 1996). Ecological economists have proposed an alternative way of viewing the relation between the economy and the environment, which is predicated not on the calculation of accounting information (Costanza et al. 1997). Drawing from the laws of thermodynamics, ecological economists maintain that the economy is a sub-system of the environment in two fundamental ways.

First, the economy depends on the environment in that it utilizes natural resources of the environment as inputs of production. Due to the laws of thermodynamics, the rate of extraction and utilization of these resources must be within the environment's ability to renew them (i.e. the environment's source function) (Costanza et al. 1997). Furthermore, because the environment is a relatively non-growing system, the scale of economic production cannot grow infinitely. It is important to note that ecological economics has emphasized the non-substitutability between man-made capital and natural resources (Daly 1996; Costanza et al. 1996). Advancement of technology, no matter how much it is, cannot substitute for natural resources. For instance, if there is no fish to harvest due to overfishing, any increase in the technical efficiency of fishing boats is a meaningless endeavor.

Second, the economy is a sub-system of the environment because economic activity is additionally constrained by the sink function of the ecosystem both quantitatively and qualitatively (Costanza et al. 1997). This emphasis recognizes that economic production generates wastes that must be transformed into harmless matters. However, due to the Entropy Law, the ability of the environment to assimilate and regenerate wastes through biogeochemical cycles is neither unlimited nor immune from human degradation. Both the quantity (scale effect) and the quality (synergistic, cumulative, and spatial effects) of economic activities can damage the sink function of the ecosystem.

The above view of the relation between the economy and the environment is fundamentally different from that of mainstream environmental economics. While mainstream environmental economics subject the environment to the economy and environmental issues to cost–benefit analyses, ecological economics understands that the economy is constrained by the environment's limited source and sink functions and that a technological transformation in consideration of the environment's biophysical constraints is bound to happen for a sustainable economy.

This is not to entirely discard the prevalent applications of cost–benefit analyses. Cost–benefit analyses are most frequently used by firms to decide "the inputs they use, the products they make, and the methods of production they utilize" to maximize profit in today's monetary production economy (Forstater 2003). However, the maximization of monetary profits is not necessarily consistent with the achievement and the maintenance of a sustainable economy in terms of environmental sustainability or, as Keynes focused on, full employment. The optimal level of pollution might be economically justifiable in a cost–benefit analysis for firms but not environmentally viable/sustainable in consideration of real biophysical limits of the ecosystem.

Hence, while environmental economics studies the economics of the environment, ecological economics studies the environment of the economy. The former assumes the superiority of monetary efficiency while the latter understands that money is merely accounting information that is worthless in the face of ecological disasters. This critical difference between environmental economics and ecological economics inspires different approaches to addressing environmental degradation. The Chinese state in recent years has moved away from hoping the EKC to self-fulfill to actively addressing environmental issues with synthesized environmental-industrial policies. This is evidenced by the Chinese state's prioritizing environmental efficiency over monetary efficiency when tackling environmental degradation. One notable example is the Chinese state's "irrational" investment in solar energy, which has been criticized by many economists as inefficient and wasteful (Naughton 2018) but celebrated by environmentalists as China's massive state subsidy in the solar industry had contributed to the 80% reduction in the cost of solar panels world-wide. In neoclassical terms, these supposedly wasteful and inefficient government interventions have in fact enormous positive externalities.

4 Contradictions in China's Environmental Policies from 1978 to 2005

Considering China's alternative economic system ("Socialist Market Economy with Chinese Characteristics"), the contradictions in China's environmental policies from 1978 to 2005 can be best illustrated via the lens of China's political economy. Between 1949 and 1978, Socialist China devised the country's first major environmental policy, "Three

Rivers and Three Lakes,"[7] in the 1950s to control floods, construct hydroelectric power, and address rural and urban water supply (Ross 1998). While the Cultural Revolution from 1966 to 1976 pushed environmental protection into the background, China's State Council started to issue environmental protection laws in 1975, with "half of these statues dealt with protecting nature and combating pollution, the rest concerned governmental administrative tasks, enforcement procedures, and the responsibilities of scientific research and propaganda agencies" (Li 2006). The environmental legislative work continued and, by the 1980s, the Chinese state government started to emphasize the importance of sustainable development—Chi Xu Fa Zhan—in the context of international environmentalism at the time, which was defined as "development that meets the needs of the present without compromising the ability of future generations to meet their own needs" (Brundtland 1987). Since the 1990s, the state has allocated significantly more funding to environmental protection in China. This was especially so during Hu Jintao's tenure of "scientific developmentalism" that saw the state "between 2001 and 2005… appropriated 360 billion yuan (US$ 45 billion) primarily for pollution control and prevention, [which is] twice as much as the total expenditures for environmental control programs in the Sixth, Seventh, and Eighth Five-Year Plans (1986–2000) combined" (Li 2006).

However, despite the above efforts in legislating environmental laws and funding environmental protection, China's environmental policy between 1978 and 2005 remained overall ineffective and often contradictory and subsidiary to its economic goals. A notable example is China's Green GDP project, which started in 2004 but was discarded a few years afterward (CAEP 2013). More importantly, China's environmental laws were becoming comprehensive, but the implementation was incomplete; China's funding for environmental protection was growing rapidly, but the scale was still insufficient, and the approach was defensive in nature (Bao 2006; CAEP 2009). Two important causes of the contradictions in China's environmental policy during this period, however, were the Chinese economy's transformation from an agricultural to an industrial economy and the Chinese state's conscious decision to prioritize economic growth over environmental concerns. As one senior NEPA (National Environmental Protection Administration) official

[7] "Three Rivers" refers to the Huai, Hai and Liao Rivers; "Three Lakes" refers to the Tai, Dianchi, and Chao lakes.

noted, "Only government regulation can successfully curb environmental degradation in a market economy" (Jahiel 1997).

As elaborated in my other paper on China's alternative economic system, China's post-1978 economic reforms have resulted in the decentralization of the Chinese economy, such as the dissolution of the Agricultural Collectives in the countryside and the restructuring of the State-Owned-Enterprises in urban China. While adopting a gradualist approach to reform, the Chinese state had increasingly granted autonomy to provincial, municipal, and local governments, state-owned and private enterprises to decide on economic matters. Deng Xiaoqin's "development is the only hard truth," therefore, "gave officials at the provincial level and below the incentive to develop their local economy, and the wherewithal to do it" (Shirk 1993). The consequence is that "most city, county, and township bureaucrats have become obsessed with encouraging economic development in their locality, especially as their political competency is evaluated based on it" (Li 2006). For example, as Boxer observes, the same construction of the thermal and hydroelectric generating plants, water reservoirs, and other energy-related systems that spurred China's economic growth has also caused population dislocation, soil erosion, increased deforestation, and a loss of arable land and natural habitats in China (Boxer 1989). As another example, the rapid industrialization in China that has led to a rise in income and higher levels of food consumption has also resulted in the alarming loss of farmland (Smil 1995).

Hence, the contradictions in China's environmental policy reflected "deep contradictions that developed in... [China's] transformation from a traditional agricultural society to a modern industrial one and Chinese national rejuvenation in international structure" (Bao 2006). In this context, the Chinese state has prioritized economic growth for most of the years after the 1978 economic reform. This was evidenced by China's refusal to sign the emissions restrictions agreement at the United Nations Convention on Climate Change in 1997 in Kyoto, Japan. As the Chinese delegate argued in the convention, "poverty eradication and economy development are still the overriding priorities of China," and that "it is not possible for the Chinese government to undertake the obligation of reducing greenhouse gases until the economy develops" (Wan 1998). This official stance has led many to predict a rather gloomy environmental future for China, as Li wrote, "in light of its vast territory and traditional economic structure, it is unlikely that China will adopt strategies that promote 'clean industries' to minimize environmental damage." Li further argues as recently as in 2006 that, "it is estimated that China will maintain its focus on heavy industries, such as petrochemicals, steel, construction materials, power generation, and coal mining, which employ technology that is more wasteful in energy and resources." It therefore appeared that, "the ethos of the reforms and the political economy constructed to support reform goals are antithetical to solving China's environmental problems" (Jahiel 1997).

Not only were environmental concerns less prioritized, from 1978 to 2005, the Chinese state has been ineffective at enforcing existing environmental laws due to its overly complex institutional arrangement that resulted in bureaucratic fragmentation. The major environmental regulatory institutions in China during the 1978–2005 era include: The National People's Congress (as the highest legislative body), the Committee of Environment and Resources, the National Environmental Protection Agency (EPA), and the National Committee of Environmental Protection in the State Council. As Bao observes, "The various organizations that are responsible for formulating, implementing and supervising environmental policy have overlapping functions and unclear rights and responsibilities" (2006). Additionally, various national ministries and committees have their separate environmental bureaus that operate independently from the national EPA.

Not surprisingly, "these bureaus generally follow the regulations of their own ministries and committees when the ideas of the national EPA and its own ministries differ" (Bao 2006). Furthermore, local environmental protection bureaus are more vertically integrated to their local governments (which determine their personnel and allocate their administrative funds) than to the national EPA. This means that "when local authorities push for high economic growth and pursue administrative goals, local environmental protection bureaus are often unable to implement environmental protection policies. In certain cases, some even become accomplices of the polluters they are supposed to oversee" (Bao 2006). Institutionally speaking, as long as promotion for provincial, municipal, city, and township-level officials is still primarily based on their ability to promote economic growth, implementation of environmental laws at the local levels will continue to be a major problem for China to address environmental degradation.

Due to China's above particularity, conventional policies such as the Pigouvian Taxes as informed by the mainstream environmental economics have seen limited application and effectiveness in China. First, China's fee and penalty systems mainly target the larger scale state-owned enterprises. However, these fees and fines were generally ineffective. They were simply augmented to product prices without affecting sales since most of these SOEs were monopolies (Bao 2006). Second, the rural township and village enterprises and private enterprises, whose competitiveness does depend on low prices, often were able to find ways to evade the fees and penalty system. Local governments, in consideration of their political competence (which is measured by the locality's economic performance), had the tendency to "race to the bottom" in order to attract business investments. Third, the discharge fee and fines are far too low to motivate enterprises to treat environmental pollution seriously. As Bao argues, "the discharge fee of noise is only 0.3% of the cost of product in a levied enterprise. The discharge fee of sewage is only 0.03% of gross production and 0.05% of the total cost of a levied enterprise. The discharge fee of SO_2 for industrial coal is only 0.20 Yuan/kg. These penalties are low especially when compared with a growing national economy and the per capita income" (Bao 2006; Xu 2011). Finally, the levying scope was narrow during this era. For example, only suspended particulates and SO_2 were levied a fee when it came to the pollution caused by burning coal while other pollutants such as CO_2 and NO_2 were excluded (Bao 2006; Xu 2011).

In conclusion, despite efforts in legislating environmental laws and funding environmental protection, China's environmental policy between 1978 and 2005 remained overall ineffective and often contradictory and subsidiary to its economic goals. These contradictions in China's environmental policy were largely due to China's transformation from a traditional agricultural society to a modern industrial one and the Chinese state's explicit prioritizing economic growth over environmental concerns. During this period, the Chinese state has been ineffective at enforcing existing environmental laws due to its complex and fragmented environmental bureaucracies and the conflict of interest that existed at the local, municipal, and provincial levels. Consequently, conventional environmental policies such as the Pigouvian Taxes as informed by the mainstream environmental economics had seen limited application and effectiveness in China. The question for the Chinese state remains: how to address the apparent contradictions between economic growth and environmental protection? How shall a socialist market economy with

Chinese characteristics engineer a green civilization while elevating the living standards of the Chinese people and actualizing the "Chinese dream"?

5 From Contradiction to Synthesis Since 2005

As discussed above, the contradiction between economic growth and environmental sustainability persisted for most of the years after China's 1978 economic reform. Since 2005, however, the Chinese state has increasingly moved away from hoping the EKC to self-fulfill with ad hoc environmental fees and fines to actively addressing environmental degradation with systemically designed and implemented environmental-industrial policies and laws. There may be many factors that had contributed to the shift of attention of the Chinese state toward environmental issues, including the increasing number of environmental protests, the growing research that shows the adverse health effects of pollution, the impact of climate change on national territory, and the urgency to ensure energy security (Chen et al. 2013; HEI 2016; Piovani 2017). From the political economy perspective, the shift of attention may be another reflection of the reform philosophy of the Chinese state—concentrate on solving the most important contradiction first (the contradiction between a low level of economic growth and the Chinese people's desire for a higher standard of living) and then tackle the problems (such as environmental degradation and income inequality) that resulted from solving that contradiction. Whatever the reason(s) might have been, China's environmental policy climate since 2005 has shifted from contradiction to synthesis that aims to create policy and legal institutions appropriate for fostering a green civilization. While it is difficult to present the totality of China's evolving environmental policy climate, I highlight two important undertakings as led by the Chinese state after 2005: improving energy efficiency and conservation for today and investing in renewable energy for the future. Both point to the Chinese state's willingness and ability to align monetary and environmental efficiencies by synthesizing crafting environmental-industrial policies and laws, which reflects China's pursuit of an alternative economic system—a "Socialist Market Economy with Chinese Characteristics."

Since 2006, the Chinese state has made substantial investments and regulatory efforts in energy efficiency and conservation to balance the need of economic growth and environmental sustainability in the short and medium terms. Given the trend of China's increasing energy demands in the foreseeable future, investments in energy efficiency and conservation are a transitory necessity to a future of renewable energy that is under development. As Zhang argues, "improving energy efficiency is considered the cheapest, fastest, and most effective way to keep energy growth under control and address environmental concerns" (2013). Considering that industry accounts for about 70% of China's total energy consumption (Zhou et al. 2010), it made sense for the Chinese state to roll out environmental-industrial policies to "encourage technical progress, strengthen pollution control, and to promote industrial upgrading and energy conservation" (Zhang 2013). In April 2006, the Chinese state established the "Top 1000 Enterprises Energy Conservation Action Program" that covered 1008 enterprises in nine key energy supply and consumption industrial subsectors. As Zhang evaluates the program,

> These enterprises each consumed at least 0.18 million tons of coal equivalent (tce) in 2004, and all together consumed 47 percent of industrial energy consumption in 2004. The program aims to save 100 million tce cumulatively during the period 2006–10 (NDRC, 2006). While there are areas that need further improvements, this program goes very much as planned as far as the energy-saving goal is concerned. In September 2011, NDRC reported that the Top-1000 Program had estimated to achieve total energy savings of 150 million tce during the 11th five-year plan period (NDRC, 2011). (2013)

Just as in other successful reforms, the Chinese state followed the "experiment, evaluate, and expand" approach in this case as well to gradually implement industrial policies to promote energy efficiency and conservation. With the lessons learned from and the success of the initial "Top-1000 Enterprise Program," the "NDRC [National Development and Reform Commission] and eleven other central government organizations in December 2011 announced the expansion of the Top-1000 Program to the 10,000 Program [that]... covered about 17,000 enterprises [that] include those industrial and transportation enterprises consuming energy of 0.10 million tce or more and other entities consuming energy of 0.05 million tce in 2010" (NDRC 2012). These enterprises cumulatively consumed more than 60% of the national total in 2010, and the program aimed to save 250 million tce during the 2011–2015 period. In 2014, the China Electricity Council reported that the total energy conservation already reached 309 million tce during the 2011–2014 period, implying that the 10,000 Enterprise Program exceeded its target by 21.13% one year ahead of the schedule.

In addition to crafting environmental-industrial policies to promote industrial energy efficiency and savings, the Chinese state has also tightened residential building efficiency standards in a gradualist and pragmatist manner,

> In 1986, the Chinese government issued the energy-saving design standard for heating in new residential buildings... This standard was revised in December 1995, requiring that new buildings be 50 percent more efficient by 2010 and 65 percent by 2020... [since 2005] China has been enforcing the energy efficiency design standard more strictly, requiring that in the very cold area and cold area all new buildings in large and middle cities before 2001, in small cities before 2003 and nationwide since 2006 onwards comply with the energy efficiency standard of 50 percent (MOHURD 2002, 2005). In northern and coastal developed areas, and in large cities, all the newly built buildings should meet the requirements of 65 percent local building energy efficiency standard (MOHURD, 2005). (Zhang 2013)

Besides regulatory efforts to improve industrial and residential energy conservation, the Chinese state has also backed the research, development, and adoption of supercritical (SC) and ultra-supercritical (USC) power generation technology to improve energy efficiency in coal-firing power plants (Duan 2016). While SC and USC do not make these coal-firing power plants "green," they generate much greater energy efficiency due to greater thermal efficiency compared to other cleaner coal technologies. The research, development, and adoption of SC and USC technologies as led by the Chinese state thus enabled more economic growth while reducing pollution significantly, as evidenced by China's rapidly falling energy use per unit of real GDP from around 250 grams of oil equivalent to produce $1 GDP (2011 PPP) in 2005 to roughly 180 grams in 2015 (which is much closer to the 120-grams OECD average) (Naughton 2018). As Naughton states,

> There has been a remarkable technological transformation of China's electricity sector: today 44% of [China's] electricity capacity uses supercritical technology, meaning very high temperature (over 705 Fahrenheit), high-pressure generators that are significantly more efficient less polluting.

> This is significantly more advanced than the U.S. fleet, which consists predominantly of older subcritical plants. For example, China has 90 ultra-supercritical units currently in operation, while the United States has precisely one. (Hart et al. 2017)

It is important to note that the abovementioned investments and regulatory efforts to improve energy efficiency and conservation were an integral part of China's massive and ambitious manufacturing upgrading plan—the Supply-Side Structural Reform (SSSR). While the SSSR was formally introduced in 2015 by Xi Jinping as a national industrial agenda, many components of it were already tested before 2015 that aimed to improve energy efficiency and address pollution. These policies included shutting down the especially polluting coal mines, converting the smaller, inefficient, and highly polluting coal mines into larger, more efficient, and less polluting ones, and reducing excess capacity in heavy industries (especially steel) that are especially environmentally damaging (Qi et al. 2016). With a combination of mandatory closures and market incentives, the Chinese state saw that "... by the end of 2010 [from 2005], the total capacity of decommissioned smaller and older units had increased to 76.8 GW (China News Net 2011), almost ten times the total capacity decommissioned during the period 2001–05," which "led the average coal consumed per unit of electricity generated to decline by 12.8 percent by 2012 relative to its 2005 levels" (CEC 2011; CEC and EDF 2012).

In addition to these highly coordinated and synthesized environmental-industrial policies (which are notably different from the contradictions we observed in the 1978–2005 period), the Chinese state has also "increased direct central government pressure on local actors" and created legal institutions to help improve energy efficiency and tackle environmental degradation (Naughton 2018). On January 1, 2018, the Chinese state implemented a new Environmental Protection Tax Law that replaced the previous ineffective pollutant discharge fee system that had been in place for the past 40 years (Shan 2016). While the impacts of the new law remain to be seen, some important changes compared to the old system are that the new law is tougher with mandatory collection of tax per unit of emissions for air, water, and solid wastes and that it aims to increase enforcement by allowing all revenue collected to stay at the local level (Shan 2016).

Meanwhile, in the transport sector, the Chinese state adjusted the excise tax for vehicles several times over the years to incentivize the purchase of energy-efficient cars. In addition to the substantial subsidies granted to EVs (Electric Vehicles), the Chinese state had "... from September 1, 2008, the rate for small cars with engines of 1.0 liter or less further decreased [from 3 percent in 1994] to 1 percent, whereas the rate for cars with engines of no less than 3 liters but no larger than 4 liters was set at 25 percent... Cars with engines of larger than 4 liter were taxed at the highest rate of 40 percent" (Zhang 2010). The result is that "China has set even more stringent fuel economy standards for its rapidly growing passenger vehicle fleet than those in Australia, Canada, California and the US, although they are less stringent than those in Japan and the EU. In the meantime, expanding Chinese cities are prioritizing public transport and are promoting efficient public transport systems" (Zhang 2010).

Above I have demonstrated the synthesis in China's post-2005 environmental-industrial policies and laws in improving energy efficiency and conservation. These synthesized environmental-industrial policies included the Top-1000 Enterprises Program, the expanded Top-10,000 Enterprises Program, the more stringent residential design standards, the investments in SC and USC technologies, the Supply-Side Structural Reform, the implementation of the new Environmental Protection Tax, etc. However, while improving energy efficiency and conservation is important and worthy, it does not alter the fact that China today is still heavily dependent on coal as the country relies on coal-burning for 2/3 of its electricity generation as of 2018. Hence, these policies, while effective, should be regarded as transitory solutions in the near term to balance China's needs for energy security, economic development, and environmental sustainability. To investigate China's environmental future, I now examine the Chinese state's developmental role in the research and development of renewable energy.

By implementing synthesized environmental-industrial policies to improve energy efficiency and conservation, since 2005 the Chinese state has positioned the renewable energy sector at the core of its national industrial planning and saw tremendous success. I first briefly discuss the results of China's renewable energy development in recent years as well as the ways by which the Chinese state has promoted that development. I then address the more important and nuanced question—is the political will of the Chinese state to continue its unparallel commitment in renewable energy sustainable? The answer to this question is critical to predicting China's (and, to great extent, the world's) renewable energy and environmental trend.

China's success in renewable energy development has been well-documented. When China passed its Renewable Energy Law in 2005, it had 112 gigawatts (GW) of installed hydroelectric capacity but almost no wind, solar, or biomass power (IRENA 2019). By 2018, China had 323 GW of installed hydroelectric capacity, ranking first in the world and more than tripling the second highest Brazil's (IRENA 2019). China's nonexistent wind energy in 2005 reached 185 GW of installed capacity in 2018, ranking first in the world and almost doubling the second highest U.S.'s at 94 GW (IRENA 2019). In terms of solar, China reached 175 GW installed capacity in 2018, ranking first in the world and more than tripling Japan's 56 GW and the U.S.'s 51 GW (IRENA 2019). Finally, in terms of total installed capacity of renewable energy sources, which include wind, solar, biomass, geothermal, hydropower, and marine energy, China had reached 695 GW as of 2018, more than doubling the second highest U.S.'s 245 GW (IRENA 2019), as shown in Fig. 1.

China's renewable energy development has effectively curbed the country's reliance on coal (see Figs. 2, 3, 4, 5 and 6). Kennedy (2013) observed that, "... renewable and nuclear power accounted for 94% of the growth in Chinese electricity generation in 2012. As a result, China met more than 9% of its primary energy demand from non-fossil sources that year..." In recent years, while coal and other fossil fuels still represent China's bulk of energy sources, their weight seemed to have plateaued while that of the renewable energy has climbed. This has led to CNREC (China National Renewable Energy Center) to project that, "China will not require gas as a bridge between coal and renewable energy" and that "the deployment of solar and wind power can increase significantly in the next 10 years... and become the core of the nation's energy system by 2050" (CNREC 2018).

The ways by which the Chinese state has promoted the above development of renewable energy in China are well-documented, too. As Gallagher (2013) summarized,

> Most of China's environmental policies are implemented to achieve targets set by the central government in its five-year plans. In the 12th Five Year Plan, for the period 2011 to 2015, the government set ambitious targets for renewables... the central government enacts some national-level policies and then assigns partial responsibility to provinces and municipalities to achieve the target... China's support for renewable energy was greatly enhanced in 2005 with passage of 'The Renewable Energy Law of the People's Republic of China.' The law created four mechanisms to

> promote renewable energy: (1) a national renewable energy target; (2) a mandatory connection and purchase policy; (3) a feed-in tariff system; and (4) a cost-sharing mechanism, including a fund for renewable energy development...

It is important to note that Gallagher's observation of China's renewable energy development raises a critical question: given that China's advancement in renewable energy has largely been dependent on the vision of the Chinese state, could it be sustained? In other words, is it possible that the Chinese state in the foreseeable future might change its stance on renewable energy and revert the country's energy usage back to fossil fuels? After all, China's economic development since 1949 has confirmed the dominant, evolving, and developmental role of the Chinese state in shaping economic and social processes (see my other paper on

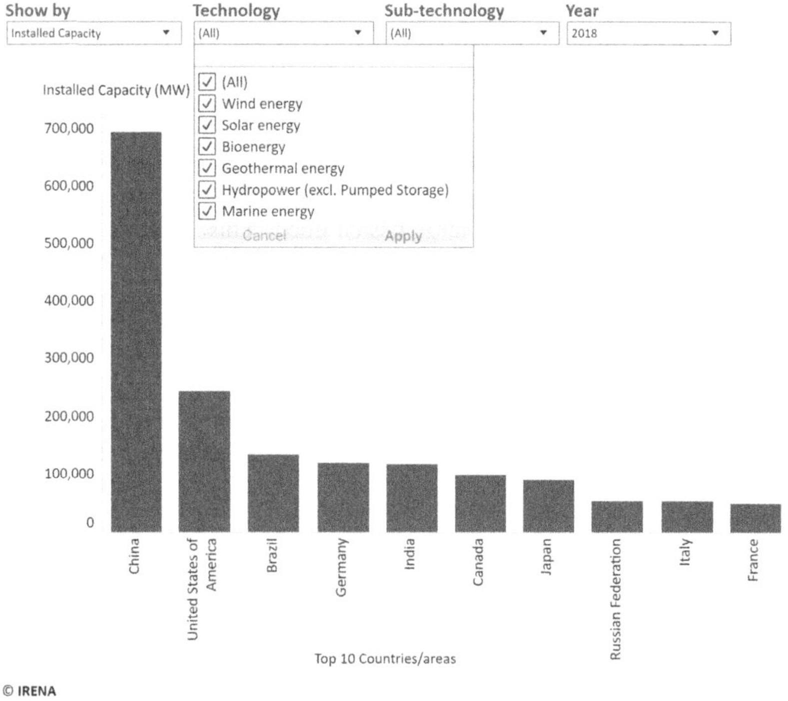

Fig. 1 Top 10 countries/areas (© IRENA)

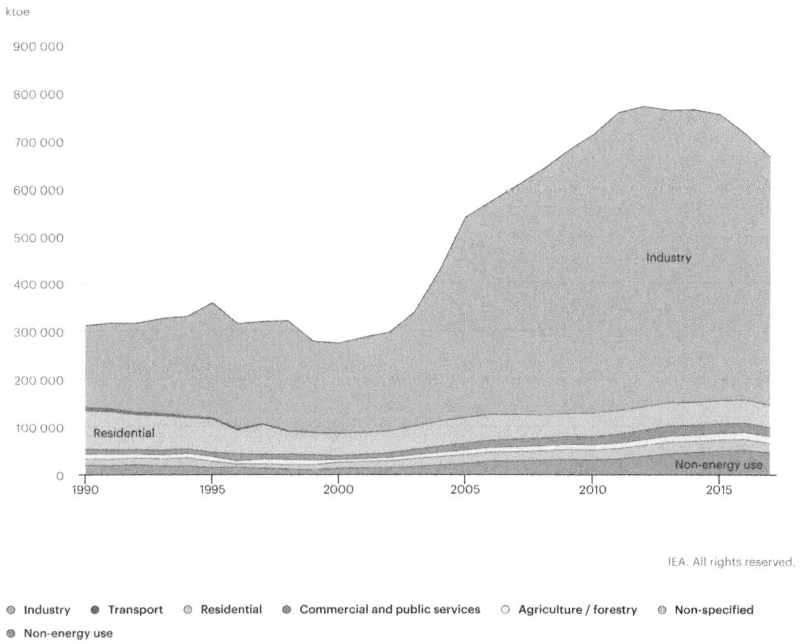

Fig. 2 Coal final consumption by sector, China (People's Republic of China and Hong Kong China) 1990–2017

China's alternative economic system). The Chinese state's vision is thus of paramount importance for us to predict China's future of renewable energy and environmental sustainability.

Though there may be setbacks (such as the recent 2018 U.S.-China trade war that saw President Trump increasing tariffs on Made-in-China solar panels), I argue that it is unlikely that the Chinese state will revert its strong commitment in renewable energy. I argue that there are three structural causes, in addition to the commonly understood ones (such as the heavy health costs associated with environmental pollution), that motivate the Chinese state to continue pursuing renewable energy development. First, from the standpoint of energy security and independence, China's main nonrenewable endowment is coal. However, "China's reserve-to-production ratio is only thirty-three years, indicating

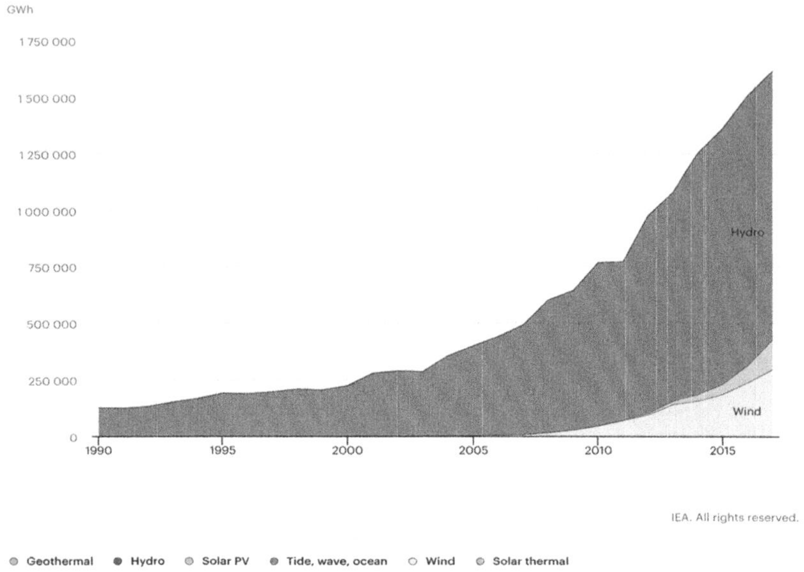

Fig. 3 Renewable electricity generation by source (non-combustible), China (People's Republic of China and Hong Kong China) 1990–2017

that China either needs to find alternatives to coal or bear the energy-security risks of becoming a massive coal importer… China has poor remaining endowments of oil and gas, quickly becoming a major importer of both. China is now the second largest oil importer in the world after the United States" (Gallagher 2011). Hence, China's resource curse may be a blessing in disguise as it mandates the Chinese state to develop renewable energy (which the country is much better endowed with) as a long-term solution to its energy security and independence while expanding foreign energy imports and investing in foreign energy production in the short run.

The second structural cause concerns with the structural change of the Chinese economy and the need for the Chinese state to find new sources of economic growth as a result. Concerning the structural change of the Chinese economy and its impact on the environment, Naughton states the following,

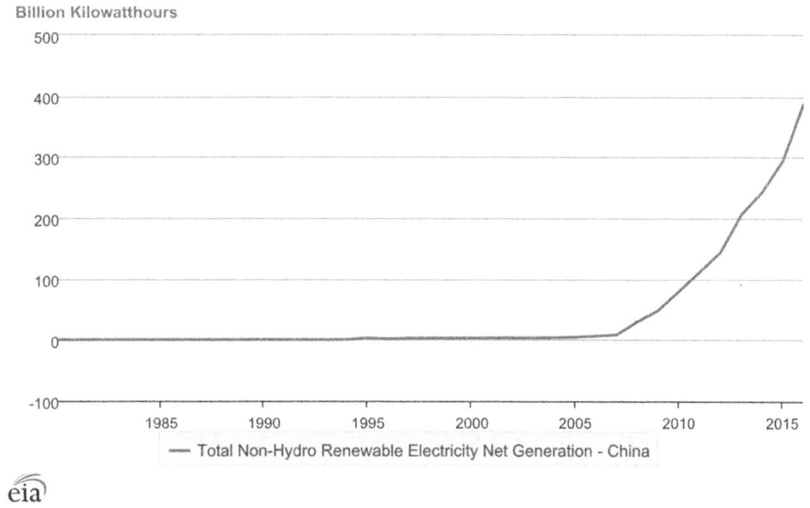

Fig. 4 Total-hydro renewable electricity net generation—China (*Source* U.S. Energy Information Administration)

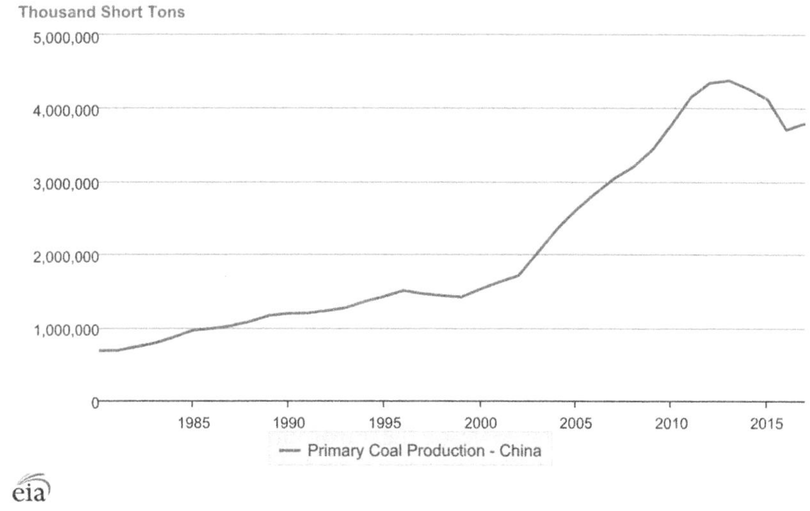

Fig. 5 Primary coal production—China (*Source* U.S. Energy Information Administration)

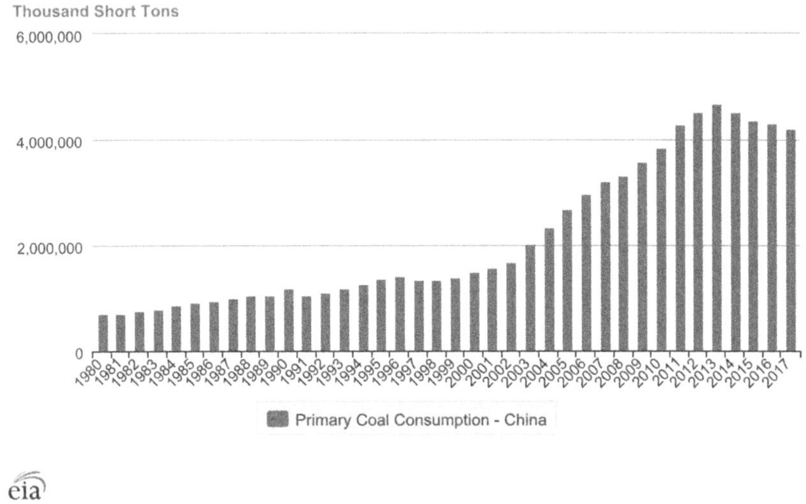

Fig. 6 Primary coal consumption–China (*Source* U.S. Energy Information Administration)

> 1. Slower economic growth: The simple fact of [China's] slower GDP growth [from double digit to around 6% in 2019] means that fewer emissions and pollutants are being released. 2. Lower investments: As growth slows, China will gradually invest a smaller share of its GDP in new fixed capital... This reduces the demand for energy-intensive industrial goods such as steel and cement. 3. Shift to services: As industrial output reaches peak levels, the growth impetus shifts to services... Most services are far less energy intensive and polluting than industry. (2018)

While Naughton's assessment may have been overly simplified, it does point out an important reason for the Chinese state to double down on renewable energy development: to find new sources of economic growth. Hence, the Chinese state's renewable energy policy has demonstrated an interesting reversal of the logic of the Environmental Kuznets Curve—instead of sacrificing the environment for economic growth, let us use renewable energy development as a vehicle to drive economic growth. As Gallagher describes this synthesis in China's economic and environmental policymaking (2013),

> In an April 2012 speech, China's Premier Wen Jiabao said, 'Greening of the economy is not a burden on growth; rather, it is an engine that drive growth and an effective means to achieve sustainable development.' As of 2011, four of the ten largest manufacturers of wind turbines were Chinese companies. Similarly, five of the top ten manufacturers of solar PV cells were headquartered in China. China's wind power industry (power generation and turbine manufacturing) generated an average of 40,000 direct jobs annually between 2006 and 2010 and is expected to generate 34,000 jobs annually between 2011 and 2020. China's renewable energy industry has reached a scale large enough to contribute directly to global price reductions, further incentivizing domestic deployment of renewable energy.

The third structural cause that motivates the Chinese state to maintain and strengthen its commitment in renewable energy development is the country's search for "indigenous innovation" (or, as Kennedy [2013] puts it, "techno-nationalism"[8]) and an early competitive edge in global strategic industries. In 2006, after several years of deliberation, the Chinese state unveiled the MLP (National Medium- and Long-Term Program for Science and Technology Development 2006–2020) to push for "indigenous innovation" and technology independence for China. As the report explains the urgency for indigenous innovation, "China's overall S&T level still has a fairly big gap to close, compared with that of developed nations… the nation will be for a long period of time under enormous pressures from developed nations [that] possess economic and S&T superiority… Facts have proved that, in areas critical to the national economy and security, core technologies cannot be purchased…" (MLP 2006). Fourteen years have passed since the release of the MLP. The rising global climate of isolationism, nationalism, populism, and protectionism since then seems to have validated the Chinese state's push for indigenous innovation that included the development of renewable energy (Lau 2018).

Specifically, the MLP listed eleven priorities that were regarded as "critical to economic and social development and national security and in dire need of science and technology" (MLP 2006). Among these priorities, developing new energy technologies was listed first and foremost. Then, in 2010, the Chinese state rolled out a more focused plan to accelerate the

[8] Adam Segal and David Kang described "techno-nationalism" as "the desire of Asian states to free themselves from dependence on western technologies."

development of seven " Strategic Emerging Industries" (SEI) in which renewable energy again occupied a strategic position. The SEI included the development of solar, wind, hydropower, nuclear, and electric vehicles, among others (Chang and Bruyninckx 2011). Since the goals of the MLP and the SEI were to push for indigenous innovation and to develop competitive advantages through early actions in global strategic industries, being part of both the MLP and the SEI entailed significant boon for China's renewable energy sector. Preferential policies were devised, such as that "China's electricity grid is obligated to purchase all the electricity generated by approved renewable energy facilities located in its service area" and that "the price for renewable energy will be set by the NDRC" (CEC 2017). Note that both policies were considered unconventional practices that distort the energy market by mainstream economists.

In conclusion, China's environmental policy climate since 2005 has generally shifted from contradiction to synthesis. The Chinese state has created policy and legal institutions appropriate for fostering a green civilization. I examined two important undertakings as led by the Chinese state since 2005: improving energy efficiency and conservation for today and investing in renewable energy for the future. I also argued that the Chinese state's determination to develop renewable energy and to guide the Chinese economy onto a sustainable path is likely to strengthen due to the abovementioned structural causes.

6 Conclusion

In this paper, I examined China's environmental reality and the evolution of China's environmental policy after the country's 1978 economic reform. I first surveyed China's dual economic and environmental crises. I then evaluated the debate between mainstream environmental economics and heterodox ecological economics regarding the relationship between the economy and the environment and the extent to which it is applicable to China. I then examined China's era of contradictory environmental policy and ineffective environmental governance from 1978 to 2005. Finally, I delineated the evolution of China's environmental policy since 2005. I concluded that in attempting to build an ecological civilization, the Chinese state's environmental policies have largely evolved from contradiction to synthesis in recent years. Furthermore, such trend should continue.

I have also demonstrated that China's "Socialist Market Economy with Chinese Characteristics" also has bearings in determining China's environmental processes. While there may be short-term setbacks and fluctuations, I remain cautiously optimistic that China will succeed in replacing most of the coal use with renewable energy by 2050 (CNREC 2018) and building an ecological civilization. While it is difficult to present the totality of China's environmental policy climate, I hope to demonstrate to readers that China's environmental policy today is in a historically stronger political and institutional position to take advantage of creative public policies to engineer environmental sustainability while achieving economic growth and full employment.

Works Consulted

Alier, Juan Martínez, and Klaus Schlüpmann. 1987. Ecological Economics: Energy, Environment, and Society. Oxford: Basil Blackwell. Print.

Bao, Maohong. 2006. "The Evolution of Environmental Policy and its Impact in the People's Republic of China." Conservation & Society, Vol. 4, No. 1 (January–March 2006), pp. 36–54. https://www.jstor.org/stable/26396647

Boxer, Baruch. 1989. "China's Environmental Prospects." Asian Survey, Vol. 29, No. 7 (July), pp. 676–677

Brundtland, G.H. 1987. Our Common Future—The Report of the World Commission on Environment and Development. Oxford: Oxford University Press, 1987, 89–95

Chang, Peifei and Hans Bruyninckx. 2011. "Wind Energy in China: From Ad hoc Projects to Strategic Policy." Renewable Energy Law and Policy Review, Vol. 2, No. 1, pp. 17–28

Chen, Yuyu, Avraham Ebenstein, Michael Greenstone, and Hongbin Li. 2013. "Evidence on the Impact of Sustained Exposure to Air Pollution on Life Expectancy from China's Huai River Policy." PNAS, Vol. 110, No. 32, pp. 12936–12941.

China Electricity Council (CEC). 2011. "Annual Development Report of China's Power Industry 2011." Beijing.

China Electricity Council (CEC). 2017. "Comprehensive Flash Report on China's Electricity Industry 2016." January 20.

China Electricity Council (CEC) and Environmental Defense Fund (EDF). 2012. "Studies on Pollution Cutting in China's Power Industry." Beijing.

China News Net. 2011. "China Decommissioned 76.825 GW of Smaller and Older Coalfired Unites During the 11th Five-Year Period." September 28, 2011

Chinese Academy for Environmental Planning (CAEP). 2009. "National Research and Pilot Project on Environmental Economics and Policy: 2009 Report." Beijing, December 30, 2009.

Chinese Academy for Environmental Planning (CAEP). 2013. "China Green National Accounting Study Report 2010." Public Version. Beijing.

CNREC. 2018 (China National Renewable Energy Center) (2018). China Renewable Energy Outlook 2018. http://boostre.cnrec.org.cn/wp-content/uploads/2018/11/China-Renewable-Energy-Outlook-2018-Folder_ENG.pdf

Costanza, Robert. 1997. "Introduction: Building Transdisciplinary Bridges at the Frontiers of Ecology and Economics." In Frontiers in Ecological Economics. Cheltenham, UK: Edward Elgar.

Costanza, Robert, Olman Segura, and Juan Martinez-Alier. 1996. "Integrated Envisioning, Analysis, and Implementation of a Sustainable and Desirable Society." In R. Costanza, O. Segura, and J. Martinez-Alier (eds.), Getting Down to Earth. Washington, DC: Island Press.

Costanza, Robert, John H. Cumberland, Herman E. Daly, Robert Goodland, and Richard B. Norgaard. 1997. An Introduction to Ecological Economics. Boca Roton, FL: St. Lucie Press.

Duan, Hongxia. 2016. "Windows of Opportunity: Coal Phase-Out in China." International Institute for Sustainable Development.

EIA (US Energy Information Administration). 2019. International Energy Statistics Forstater, Mathew. 2003. "Public Employment and Environmental Sustainability." Journal of Post Keynesian Economics, Vol. 25, No. 3 (Spring).

Gallagher, Kelly. 2013. "Why & How Governments Support Renewable Energy, Daedalus." The Alternative Energy Future, Vol. 142, No. 1 (Winter), pp. 59–77

Hart, Melanie, Luke Bassett, and Blaine Johnson. 2017. "Everything You Think You Know About Coal in China Is Wrong." Center for American Progress, May 15, 2017

Health Effects Institutes (HEI). 2016. "Burden of Disease Attributable to Coal-Burning and Other Major Sources of Air Pollution in China." Special Report #20. Boston: Health Effects Institute.

Jahiel, Abigail. 1997. "The Contradictory Impact of Reform on Environmental Protection in China." China Quarterly, Vol. 149 (March), p. 93.

Kennedy, Andrew. 2013. "China's Search for Renewable Energy: Pragmatic Techno-nationalism." Asian Survey, Vol. 53, No. 5 (September/October 2013), pp. 909–930, Published by: University of California Press. Accessed at: https://www.jstor.org/stable/10.1525/as.2013.53.5.909

Lardy, Nicholas. 1987. "The First Five Year Plan, 1953–1957." The Cambridge History of China, vol. 14. New York: Cambridge University Press.

Li, Xiaofan. 2006. "Environmental Concerns in China: Problems, Policies, and Global Implications." International Social Science Review, Vol. 81, No. 1/2 (2006), pp. 43–57

McCarthy, Niall. 2018. "Air Pollution Contributed to More Than 6 Million Deaths In 2016." Forbes.com. Accessed at: https://www.forbes.com/sites/niallmccarthy/2018/04/18/air-pollution-contributed-to-more-than-6-million-deaths-in-2016-infographic/#15fba9cf13b4

Ministry of Housing and Urban-Rural Development of China (MOHURD). 2002. "A Circular on the 10th Five-Year Outline of Energy-Saving Plan for the Building Sector." June 20, 2002.

Ministry of Housing and Urban-Rural Development of China (MOHURD). 2005. "A Circular on Requiring New Residential Buildings to Comply with Energy-Saving Design Standard." April 15, 2005.

National Development and Reform Commission (NDRC). 2006. "The Top 1000 Enterprises Energy Conservation Action Program." Beijing, April 7, 2006

National Development and Reform Commission (NDRC). 2011. "A Circular on Improving on Grid Feed-in Tariffs for Solar PV." Beijing, July 24, 2011

National Development and Reform Commission (NDRC). 2012. "List and Energy-Saving Targets of the Ten Thousand Enterprises Committed to Energy-Saving and Low- Carbon Activities." May 12, 2012.

National Development and Reform Commission (NDRC) and Eleven Other Central Government Organizations. 2011. "A Circular on Implementation Option for the Ten Thousand Enterprises Committed to Energy-Saving and Low- Carbon Activities." December 7 2011.

National People's Congress. 2005. "The Renewable Energy Law of the People's Republic of China." 2005. Accessed at: http://www.china.org.cn

Ong, Lynette. 2012. "The Apparent 'Paradox' in China's Climate Policies." Asian Survey, Vol. 52, No. 6 (November/December 2012), pp. 1138–1160. https://www.jstor.org/stable/10.1525/as.2012.52.6.1138

Piovani, Chiara. 2017. "The 'Greening' of China: Progress, Limitations, and Contradictions." Journal of Contemporary Asia, Vol. 47, No. 1.

Ross, Lester. 1998. "China: Environmental Protection, Domestic Policy Trends, Patterns of Participation in Regimes and Compliance with International Norms." China Quarterly, Vol. 156 (December), pp. 809–835.

Shan, Yuxiao. 2016. "The Environmental Protection Tax Will Be the 18th Tax Types on January 1, 2018." Caixin. December 25, 2016.

Shirk, Susan. 1993. The Logic of Economic Reform in China. Berkeley, CA: University of California Press.

Smil, Vaclav. 1995. "Who Will Feed China?" In Rozelle et al. (eds), The Impact of Environment Degradation in Grain, 801–813.

Stern, David. 2004. "The Rise and Fall of the Environmental Kuznets Curve." World Development, Vol. 32, No. 8, pp. 1419–1439

The State Council (2012). "The National 12th Five-year Development Plan for Strategic Emerging Industries." July 20, 2012

Wan, Ming. 1998. "China's Economic Growth and the Environment in the Asia-Pacific Region." Asian Survey, Vol. 38, No. 4 (April), pp. 376–378.

Xu, Y. 2011. "Improvements in the Operation of SO2 Scrubbers in China's Coal Power Plants." Environmental Science & Technology, Vol. 45. No. 2, pp. 380–385.

Ye Qi, Nicholas Stern, Tong Wu, Jiaqi Lu, and Fergus Green. 2016. "China's Post-Coal Growth." Nature Geoscience, Vol. 9 (August).

Zhang, Zhongxiang, "Energy and Environmental Issues and Policy in China." Fondazione Eni Enrico Mattei (FEEM) (2013). Accessed at: https://www.jstor.org/stable/resrep00994

Zhou, N., M.D. Levine, and L. Price. 2010. "Overview of Current Energy-Efficiency Policies in China." Energy Policy, Vol. 38, pp. 6439–6452.

Index

N. Karagiannis and J. E. King (eds.), *Visions and Strategies for a Sustainable Economy*, Global Institute for Sustainable Prosperity,
https://doi.org/10.1007/978-3-031-06493-7

The manufacturer's authorised representative in the EU is Springer Nature Customer Service Centre GmbH, Europaplatz 3, 69115 Heidelberg, Germany. If you have any concerns regarding our products, please contact ProductSafety@springernature.com

Printed and bound by CPI Group (UK) Ltd, Croydon, CR0 4YY
30/09/2026
02225490-0011